D0467432

Understanding Technology

Understanding

Technology

Charles Susskind

The Johns Hopkins University Press
BALTIMORE AND LONDON

Manufactured in the United States of America

The Johns Hopkins University Press, Baltimore, Maryland 21218
The Johns Hopkins University Press Ltd., London

Library of Congress Catalog Card Number 72-12344
ISBN 0-8018-1304-2

Library of Congress Cataloging in Publication data will be found on the last printed page of this book.

Contents

Preface ix

1 Birth of Modern Technology 1
Technology 1
History of Technology 2
Industrial Revolution 2
Takeoff into Self-sustained Growth 5
Abuses of the Industrial Revolution 6
Mature Industrialization 7
Political Consequences of the Industrial Revolution 9
Transformation of Society by the Industrial Revolution 13

2 Coming of Age of Technology 15
Continuing Evolution of Industrialization 15
Energy Conversion 17
Materials Processing 19
Transport 20
Communications Technology 23
Building Technology 24

Mass Production 27
Higher Technical Education 30

3 Rise of Modern Technology 33
The Second Industrial Revolution 33
Contributions of Electronics to the Second Industrial Revolution 35
Industrial Electronics 37
Radar 40

4 The Computer Technology 43
Cybernetics 43
Solid-State Electronics 44
Electronic Computation 48
Electronic Control Involving Computers 55

5 Some Aspects of Contemporary Technology 59
Energy 59
The New Food Revolution 62
Materials Made to Order 65
Technology and the Healing Arts 68
Technology and the Fine Arts 70
Technology and the Pedagogic Arts 74
Technology and Humanistic Studies 77

6 Ideologies of Technology 79
Technology and a Social Order Based on Humanism 79
Technocracy 82
The Managerial Revolution *by James Burnham* 83
The New Industrial State *by John Kenneth Galbraith* 85
The Technological Society *by Jacques Ellul* 89
The Quest for Utopia 93
Marxist Views of Technology 97

7 Technology as a Social Force and Ethical Problem 103
Technologist: Benefactor or Monster? 103
Agricultural Extension 105
Euthanasia 108
Division of Responsibility 113
An Engineer's Hippocratic Oath 118

8 Challenges 119
International Understanding 119
Overpopulation 124
Leisure 125
Technology Assessment 127
Alienation 132

Notes 135
Subject Index 153
Name Index 159

Preface

A curious deficiency besets our society. Educated and well-read citizens, no matter how specialized their education and reading, know a good deal about traditional culture. In America, even engineering students must complete a prescribed number of college courses in socio-humanistic studies. In Russia, the most derisive name to call a boor is *nekulturnyi* (uncultured). Yet in a "liberal" education, technology is virtually ignored, despite its central place in contemporary culture—more important in many ways in shaping human lives than the prevailing political system.

The present work will give the reader who seeks to remedy this deficiency a first acquaintance with contemporary technology. It is not a collection of descriptions of its various parts, an approach that tends to be rather dull except for readers already professionally committed to the subject. Rather, the book seeks to provide an overview of the development of modern technology and of its main social and political consequences. Some recent works by political scientists, economists, sociologists, and other specialists that touch on technology in important ways are discussed, in a few cases in considerable detail, in the hope of whetting the reader's appetite and encouraging him to tackle the originals. Only one chapter contains a descriptive account of some aspects of contemporary technology, but even there an effort is made to discuss topics that are seldom treated in that context, such as food, medicine, education, and the arts. Nor has much been said about two subjects that have had more than their share of contemporary attention: resources and the environment.

In recent years, about one bestseller a year has drawn our attention to some technological adversity: Marshall McLuhan and Quentin Fiore's *The Medium Is the Massage* (1967), Ralph Nader's *Unsafe at Any Speed* (1968), Theodore Roszak's *The Making of a Counter-Culture* (1969), Charles Reich's *The Greening of America* (1970), Barry Commoner's *The Closing Circle* (1971), The Club of Rome's *Limits to Growth* (1972). A common theme is to blame technology (and by implication, the technologist) for society's shortcomings. Perhaps society itself should be held principally responsible. More important still, all segments of society should participate in deciding how technology might be used to solve contemporary problems. But before they can participate, educated citizens first need to know something about how modern technology came about and what its effects on society are.

* * *

If my attempt to impart an international flavor to the work has been successful, it is largely the result of a pleasant year spent at the Graduate Institute of International Studies, University of Geneva, under the terms of a National Science Foundation fellowship. My debt to both sponsors is great. I am also indebted to a number of colleagues, at Berkeley and elsewhere, whose comments on the manuscript are reflected in this volume.

C.S.

Berkeley, California

Understanding Technology

Fly-ball governor rotating about vertical axis swings outward with increasing speed, reducing steam supply and hence speed. This early application of the *feedback principle* (self-adjusting windmill vanes were another) was adapted to steam engines by James Watt from a similar device first used in mills to keep the gap between grindstones constant. (R. Stewart, *Steam Engine,* 1824)

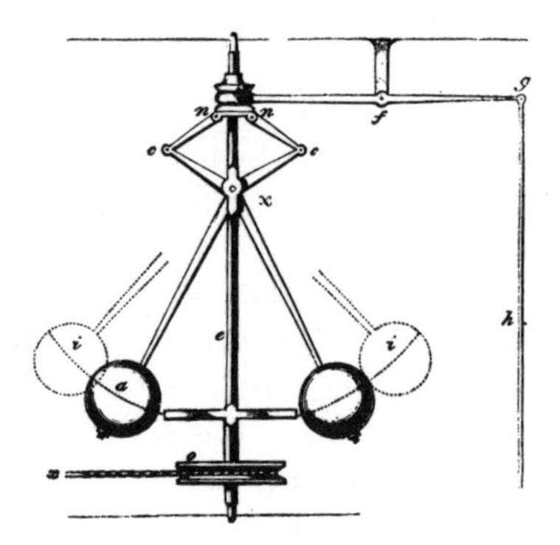

Chapter 1 Birth of Modern Technology

TECHNOLOGY, *man's efforts to satisfy his material wants by working on physical objects,* goes back to his very beginnings. Anthropologists date the transition of our prehuman ancestor into man from the time he began to use weapons and tools.[1] All that has gone since in the development of material culture—in how we do things—is technology. Its history is inextricably bound up with our own present-day culture. Elements of the early human accomplishments coexist in our contemporary world with the most recent developments. Equalization of the technological levels of various parts of the world is seen as one of our most pressing problems.[2]

Yet this book is not about the history of technology, a well-established discipline in its own right, with all that any self-respecting scholarly field can boast: professional societies, learned journals, professorial chairs, and weighty reference works.[3] In the opening chapters, we shall mention some of the major events in the history of modern technology, to provide a suitable background or *mise en scène* for the emergence of contemporary technology—but that is all. For the rest, we shall focus our attention on the technology of today, on its present goals, and on other possible goals.

HISTORY OF TECHNOLOGY, *the systematic study of past cultural, economic, and social implications of technology*, permeates all historical studies. Divisions of historic time—Stone Age, Bronze Age, Iron Age—are named after man's implements in those eras. The passage from a primitive to a civilized condition means passing from a state in which history is determined by nature to one in which man's will is the principal determinant. Instances in which technology was a determining factor abound in the history of civilization. They begin to predominate in Western civilization after the mid-18th century; we speak of that transition as the Industrial Revolution, when the lands bordering on the North Atlantic began to outpace all others in material accomplishments. There have been other transitions of comparable significance. Domestication and harnessing of animals, the rise of agriculture, development of cities and of transport across land and sea, and the invention of writing each had far-reaching consequences. The history of human accomplishment is highlighted by such "events" as the development of fortifications and other massive structures during the Greco-Roman age, the expansion of agriculture following the discovery of new ways of plowing in the Middle Ages, and the large spurt experienced by many crafts—such as printing—during the Renaissance.

None of these advances was really an event in the usual sense of something that happens at some instant. In technology, major advances are rarely signalled by someone leaping from his bath shouting "Eureka!" or being hit on the head by a falling apple. Rather, innovations are made gradually, once the necessary conditions have come into being, often independently but almost simultaneously in several places. The massive changes that follow when technological innovations are introduced into a suitable environment transform the entire society, not merely its technology, though political historians have been slow to acknowledge that. In the history of Western civilization, we speak of the "Middle Ages" (the millennium that followed the disintegration of the Western Roman Empire), and we call the first half the "Dark Ages" because of the general decline of such cultural institutions as political organization and art and literature. Yet we now know that in most of Europe technology (and with it the common man's lot) continued to advance during that period, and the Dark Ages no longer seem quite as dim as they used to.

INDUSTRIAL REVOLUTION, *the shift of a substantial part of a primarily agricultural work force living on the land to factory manufacture centered in towns,* was an "event" in the sense mentioned, the culmination of a continuing technological evolution that was at least a century and a half in the making. When we speak of *the* Industrial Revolution, we mean a specific period in the history of Britain, 1750–1830. The term was first used by Prosper de Launay, a French deputy, in a parliamentary debate in 1829, and taken up later and popularized by the first Arnold Toynbee (1852–1883). Industrial revolutions in other countries followed. France and Germany came next, then other nations. The USA participated from the first—Eli Whitney (1765–1825) invented the cotton gin in 1793 and began manufacture by assembly from interchangeable parts in 1798—but an industrial revolution on the British model did not take place in America until after the Civil War of 1861–1865. In Japan and in Russia, the industrial revolution did not begin until about 1900; in some other non-European countries, it was even later; and in still others it has yet to come. We turn to the British model, and to its antecedents and consequences, to see what useful insights we might gain for our study of contemporary technology.[4]

The Industrial Revolution, based on the water wheel, is epitomized by the development of the factory system and—following the invention of the steam engine—of a heavy capital-goods industry. Yet in a broader sense these are mere episodes in a complicated process that began in England in about 1600, during the astonishing flowering of artistic and intellectual life known as the Elizabethan Age. Francis Bacon (1561–1626) characterized knowledge as "a rich storehouse for the glory of the Creator and the relief of man's estate." The idea that technological innovations represented progress, coupled with the vision that Bacon and other utopian writers had of an ideal society, free from poverty and misery, played an important part in Western history. By and large, European religious attitudes (and among them the Protestant rather more than the Catholic) were more favorable to material progress than those in the East.[5] Scientific societies were founded on the premise that they would enhance material prosperity, even though we cannot agree with the contention of men such as Count Rumford (1753–1814) that "invention seems to be particularly the province of men of science" unless we broaden that category to include craftsmen and others whose interests were more practical than scientific. But—progress in agriculture apart—the outstanding feature of the 150 years that preceded the Industrial Revolution was the increase in commerce following the great voyages of exploration, a growth that led to the accumulation of the savings without which no other material growth is possible.

From overseas colonies came gold and silver, which increased the supply of ready money and led to an expansion of banking and of the use of credit. New agricultural products such as the potato, sugar cane, tobacco, and cotton were introduced just as European agriculture was undergoing certain improvements in methods of farming and stockbreeding. With the land capable of feeding more people, population increased, cities grew, and demand for all consumer goods (including food) encouraged more production—as did the appearance of markets represented by colonial populations such as those in America and other underdeveloped lands under European control. The realization that material goods and services available to a nation were a better measure of its wealth than its bullion holdings led to the gradual decline of mercantilism, an economic theory that happened to favor exploitation of the colonies for the benefit of the motherland. The American Revolution and similar happenings elsewhere were to a considerable degree expressions of the colonials' distaste for the latter arrangement.

Not all the developments that took place in Europe during the period 1600–1750 can be traced to the expansion of trade, important though that was. It is likely that contemporary improvements in the cultivation of crops and in animal husbandry, for instance, would have come even without external stimulus. Much progress was made in mining and in the processing of metals, especially iron. Simple machine tools—for example, lathes for screw cutting and lens grinding—were developed and used to make quite complex devices such as telescopes, microscopes, clocks, and other precision instruments. Transportation and building achieved new heights. It is difficult to assess the role that the flowering of commerce played in these advances. It is probably safe to say that none of them remained wholly untouched by the aura of prosperity and social and economic opportunity that characterized Britain in 1750.

Why Britain? France and Germany were both larger and richer. The populations of Austria, Italy, the Netherlands, or Sweden certainly contained talented individuals in similar proportions and boasted living standards no lower than the British. But France was saddled with political and financial systems that left little capital available for investment. Germany was an untidy assemblage of quasi-independent states beset, like their neighbors, by debilitating conflicts ranging in extent from the knock-down, drag-out Thirty Years' War (1618–1648) to more localized donnybrooks. Only in Britain did the right "mix" of conditions favorable to industrialization arise: capital savings and willingness to use them, raw materials, adequate labor supply, preexisting high levels of craftsmanship and social organization, growing markets, and public and private attitudes favorable to innovation.

TAKEOFF INTO SELF-SUSTAINED GROWTH, *the continued multiplication of goods, labor, and services that characterizes the beginning of any successful industrialization,* thus made its debut in Britain. The popular symbols—the steam engine and the railroad—actually belong to a more mature stage. In the beginning, there was cotton, an industry in which rapid expansion could be based on relatively modest financial and technical resources; and a considerable colonial market over which the British held a virtual monopoly. There had never been anything like it. Weaving was mechanized by the invention of Kay's flying shuttle (1733), Cartwright's power loom (1785), and Jacquard's draw loom for weaving patterns (1804). Spinning, which like weaving had been accomplished largely by hand since antiquity, was also mechanized throughout this period, in part to feed the voracious appetite of the new mechanical looms, although cause and effect in the contemporaneous improvements in weaving and spinning cannot always be clearly discerned. Wyatt and Paul patented a machine for drawing, twisting, and winding yarn in 1738; Hargreaves invented the jenny, a frame for spinning several yarns at once, in 1765; Arkwright patented a stronger, improved frame in 1769; and in 1779, Crompton combined several features of these machines into a hybrid device, the mule, adaptable to water (and ultimately steam) power.[6] The harvesting of cotton was greatly facilitated (and reduced in cost) by the invention of Whitney's gin, a sawlike machine for separating cotton fiber from the seeds, in 1793. All these improvements could be introduced without much capital outlay. Robert Owen (1771–1858), one of the more successful capitalists of the era before he got religion and became one of the founders of the cooperative movement, borrowed a hundred pounds to start his first mill in 1789; twenty years later, he took over a textile firm by paying off his partners the sum of £84,000 in cash (although marrying the boss's daughter also helped). The enormous profits, out of which new machines typically paid for themselves in a year or two and thus provided unequalled incentives for innovation, could obviously not last forever, since competition among manufacturers gradually reduced the margin of profit; but the total market continued to expand astronomically, and even small profit rates produced large fortunes.

The expansion of British cotton manufacture would have been fast even if Britain had not monopolized the world markets, but since she had them virtually to herself, the growth was precipitous. By 1785, annual British cotton exports had grown to the incredible yearly volume of 40 million yards of cloth. By 1850, exports had again increased more than 50 times, to a fantastic 2025 million yards a year. Britannia ruled the waves, and with them the markets. Cut off from other European countries by the blockades of the Napoleonic wars, entire subcontinents such as South America came under British economic sway, which remained unbroken for over a century. In India, a promising native cotton industry was actually forced out of existence by the mercantilistic policies of the mother country. (British textiles entered India duty free, but Indian manufactures were barred from Britain by high tariffs.) In Egypt, where an ambitious and enlightened Turkish governor, Mohammed Ali (1769?–1849), had tried to follow the end of Napoleon's occupation in 1801 by industrializing the country, modernization was nipped in the bud by policies Britain and France dictated to the moribund Ottoman Empire.[7]

Nor was the supply of the raw material held back by competition or by reliance on the sluggish European agriculture. First in the West Indies and then in the southern United States, vast plantations run on a pattern quite different from the European, manned by slave labor and unconcerned with conserving soil, managed to keep up with the ever-increasing appetite of the Lancashire mills. (The planters in turn became fairly heavy buyers of the finished cotton goods.) Thus political dominance, monopoly of the market, unrestricted supply of raw materials, and technological innovation combined to impart a ground speed to the British cotton industry fast enough to make it take off—in fact to let the entire economy become airborne under its own power. But it was not to be a smooth flight.

ABUSES OF THE INDUSTRIAL REVOLUTION—*exploitation of labor, headlong urbanization, the uncontrollable trade cycles of boom and slump*—became apparent quite early in the game. On the average, the laboring classes were no worse off then their peasant fathers and brothers. (At any rate, there was never any shortage of recruits from the land for the factory.) But the dislocation of industrialization produced severe regional hardships extending over periods long enough to lead some of the best thinkers of the age—

men such as Thomas Robert Malthus (1766–1834) and David Ricardo (1772–1823)—to the erroneous conclusion that the economy had no way to go but down. Mechanical innovations such as the power loom threw thousands of hand-loom weavers out of work; many literally starved to death. Other workers led a hand-to-mouth existence in the squalid slums, exploited, ill fed, pauperized, demoralized, prey to frequent epidemics. Drunkenness, crime, prostitution, and other vices were widespread. The new proletariat was largely urban—or rather suburban: the later separation of large European towns into a middle-class, residential West End and working-class industrial suburbs to the east, based on the prevailing westerly winds, goes back to this era. The greatest threat to security was the fact that the industrial economy, despite a continuing upward trend, was no more exempt from recurring depressions than had been the simpler, largely agrarian economy that had gone before. Apart from occasional business failures of the weakest firms, the brunt of this instability fell on the workers, who had no money to tide them over hard times and not much other recourse, private or public. Even when the cotton business boomed, profits were under continuous pressure from competition. The fear that profits would be reduced to a point at which few investors would appear and the economy would be reduced to a "stationary state" haunted economists from the time of Adam Smith (1723–1790) right up to John Stuart Mill (1806–1873). The price of the raw material had been pushed down as far as it would go; mechanization was proceeding apace; and the only costs that could still be cut were wages. Before long, they were near starvation levels, and there they remained, until the better and cheaper distribution of food that followed the introduction of railroads and steamers pushed down the cost of living. As we shall see, the advent of railroads actually solved most of Britain's other economic problems as well.

MATURE INDUSTRIALIZATION *is characterized by adequate capacity for producing capital goods,* above all iron and steel. The Industrial Revolution had turned Britain from an almost exclusively agricultural economy into one dominated by industry. But it was overwhelmingly a single industry, textiles, and its machines were run by animal or water power. Production of iron and iron products was modest, about 250,000 tons in 1800, and less than that after the end of the Napoleonic wars brought a drop in military demand. There was one other British industry, as yet little mechanized, of considerable size: coal mining, whose output in 1800 ran to 10 million tons, nearly nine-tenths of the world's total. It was the coal industry that found use for the first steam engines, both in mining and in conveying coal. And when the conveyors had grown into the first railroads, they in turn became the principal users of iron, nudging that industry into self-sustaining growth and at the same time providing the new cotton-rich bourgeoisie with domestic investment opportunities.

A primitive engine using steam for pumping water from mines was patented by Thomas Savery (1650?–1765) as early as 1698, but it only worked to heights of a few meters. A much improved mechanical steam pump, invented in 1712 by Thomas Newcomen (1663–1729), was phenomenally successful and had the field virtually to itself for sixty years. Beginning in 1765, James Watt (1736–1819) made a number of improvements that greatly raised the efficiency of the steam engine and extended its uses beyond mere pumping. Chief among the improvements were the separation of the hot steam cylinder from the cooler condenser, a linkage to convert the reciprocating (up and down) motion to rotation of a wheel, and utilization of the fly-ball governor for controlling engine speed. (This last, the familiar pair of rotating weights so suspended that they swing out centrifugally with increasing speed and thereby cause a compensating motion of the valve that controls the supply of steam, was also one of the first applications of the *feedback* principle so important to all control systems.)

Watt's steam engines were a great success. Hundreds of them were in use by 1800, gradually replacing (though not wholly supplanting) the earlier Savery and Newcomen engines. They not only took the place of water wheels and wheels turned by horsepower (in terms of which the output of the new engines was measured) but also made entirely new extensions and concentrations of power possible: deeper mines, centralized factories whose location no longer depended on water power, and all the regularity and discipline that henceforth characterized working conditions away from the worker's home. But an even greater change was in the offing: the high-pressure steam engine, whose smaller weight and size had to be achieved before steam power could be applied to transport vehicles and vessels. Such engines were introduced almost simultaneously around 1800 by Richard Trevithick (1771–1833) in England and Oliver Evans (1755–1819) in America. Leaving aside numerous predecessors and rivals, we may say that the first commercially viable steamboat was the *Clermont* of John Fulton (1765–1815), which began service in New York in 1807, and that the first reliable locomotive was the *Rocket* designed by George Stephenson (1781–1848), his son Robert Stephenson (1803–1859), and Henry Booth (1788–1869), whose design bested all others in the famous trials held at Rainhill near Liverpool in 1829.

Railroads consisting of carriages drawn along wooden rails had been thought of before, and even iron rails were in use. The first European passenger road of this sort, between Budweis in Bohemia and Linz in Upper Austria, had been devised (to link the Moldau-Elbe river system with the Danube basin) as early as 1807 and put into operation in 1827. Goods were also hauled along rails by systems of stationary engines and ropes, mainly over short distances, as in taking coal from mine to waterway. Thus, once the feasibility of a train of carriages drawn by a locomotive had been shown, the system was quickly adopted in all countries that had any sort of industrial capacity at all. And the rails were iron—in fact the word for railroad is "iron road" (*chemin de fer, Eisenbahn, ferrovia, zheleznaya doroga*) in most languages except English. The railroad became far and away the greatest user of iron (and hence of coal, which is used in smelting iron ore), creating the first large-scale capital-goods industry in country after country, beginning in Britain, where a rail network was elaborated soonest. The total plant—the shops, bridges, earthworks, stations, and other structures—also sopped up excess capital like a sponge, despite the fact that returns were not particularly large. For years, British investors had financed one shaky foreign government and enterprise after another—and had lost their money more often than not. They turned to railroads with enthusiasm: here was an investment opportunity at last that fired the imagination and that produced tangible goods, first at home and later increasingly overseas. Most of the railroads in North and South America and a good part of those on the Continent were financed by British capital. If they yielded less than 4 percent—what of it? They were the future. They could haul passengers and goods in unprecedented quantities, at speeds that represented the first substantial increase in 2000 years, opening up entire continents and linking sites of raw material with ports—a surefire speculation. By 1850, British investors had sunk £240 million in domestic railroads alone—and had brought industrialization to maturity.

POLITICAL CONSEQUENCES OF THE INDUSTRIAL REVOLUTION *were dominated by the unequal distribution of its bene-*

fits. No modernization is possible without capital savings, and the inability to accumulate capital is what holds back preindustrial economies to the present day. The questions of how to go about accumulating such savings and whether the British could have done it without producing quite so much misery are full of political overtones. On the one hand—the *left* hand—we are told that a rationally ordered society can certainly industrialize and improve its living standards simultaneously. On the other hand, say the champions of unfettered individual enterprise, not so fast; the Marxist way leads to regimentation and authoritarianism, and the rise in living standards in countries that have followed it has lagged far behind that in capitalist societies. Partisan interpretations of history are, of course, of no help in trying to decide what could have been. The Britain of 1800 could not have been expected to know how to curb the flight of the strange bird it had launched, industry, even supposing that curbing had been conceived as desirable; and the doctrines of Marx and his predecessors were the *results* of the Industrial Revolution, not preconditions.

There can be little doubt that the laboring masses were made to carry an unfair portion of the costs of industrialization. But the conditions under which they worked and lived must be seen against a background of the times, not in terms of highly industrialized economies. We know now that, in a society in which unemployment is widespread before industrialization, its introduction cannot immediately and by itself cure all ills. On the contrary, low-cost labor retards industrialization and postpones its benefits: given a large pool of labor, wages are likely to remain low, working conditions shocking, and the necessities of life expensive. If at the same time the population is growing in size, the society is well launched toward catastrophe. Under such conditions, industrialization may in fact be understood as the means of barely averting catastrophe rather than as a panacea. Without industrialization, wholesale starvation may overtake entire provinces, as it did in the Irish famine of 1847, when about one-eighth of the population starved to death and a like portion emigrated following a disastrous failure of the potato crop.

We also know that the costs of industrialization can be more equitably distributed than they were the first time around. The industrial workers of early 19th century Britain, though no worse off than the landless agricultural laborers, lived under appalling conditions. The poets saw it first. Surveying "England's green & pleasant land," William Blake (1757–1827) asked fiercely, "And was

Jerusalem builded here / Among these dark Satanic mills?" Some of the hand-loom weavers, facing increasing competition and unemployment, turned to direct action of the only sort they could think of: smashing machinery. Beginning in Nottingham toward the end of 1811, under the leadership of a mysterious (and possibly nonexistent) "Ned Ludd," bands of workmen roamed the English countryside at night, entering the smaller factories and breaking or burning down the mechanized wooden looms. The Luddites received quite a bit of popular support, at least as long as they avoided bloodshed.[8] But presently their activities grew to riot proportions, troops were called out, an employer who had requested protection was found murdered, and the movement was brutally suppressed, with several hangings and transportations, by a government still nervous about the sailors' mutiny at Spithead fifteen years before. (The plot of Herman Melville's novelette *Billy Budd, Foretopman*, a realistic account of conditions aboard a British warship of the day, takes place in the same period of unrest.)

As the 19th century began, Britain was one of a tiny number of constitutional democracies (with Switzerland, the young USA, and —intermittently—France) in a world in which all other functioning states were absolutist monarchies, slavery and serfdom were widespread, and suffrage was limited to a few property owners even in the democracies. Civic organization did not keep pace with urbanization: an industrial center such as Manchester, citadel of the British textile industry, remained virtually unrepresented in Parliament until the Reform Act of 1832, when the town became a parliamentary borough and eligible for two members of Parliament. (Manchester was administratively made up of six townships that were incorporated into a single borough in 1838, but not before the population had grown to a quarter million.) Parliament was dominated by members of an increasingly irrelevant landed aristocracy, who imposed the Corn Laws on the country—import duties that served to keep food prices high after the Napoleonic wars ended in 1815—and kept them in force until a pressure group of free-trade manufacturers forced their repeal in 1846, in the name of laissez faire. But the laissez-faire doctrine also prevented effective control of working conditions in the factories. Apart from legislation to protect the health, safety, and morals of children (and later of women) laboring in the textile mills, the law did very little for the workers. Agitation for political reform, which antedated both the American and the French revolutions, was forced to lie low during the protracted "emergency" of 1789–1815. Finally, in 1832 the reform movement, drawing strength partly from the workers and partly from the general disgust with the corruption that was the consequence of restricting the vote to 400,000 out of a population of 24,000,000, managed to have the electorate somewhat enlarged by the Reform Act (it was tripled) and the representation redistributed—and there it stayed, except for another increase and redistribution in 1867, for over 50 years, still favoring the rural areas over the towns. For the political power of the urbanized workers remained at best indirect, their leadership pitiable, their very ideology uncertain and ill defined.

This ideology derived vaguely from the *Enlightenment*, the 18th-century fountainhead of all liberal thinking on both sides of the Atlantic: the belief that the rational, secular approach that had been so spectacularly successful in science would also serve to solve all problems of society. This belief might lead one to conclude that enlightened self-interest (the "pursuit of happiness" of the American Declaration of Independence) would be all the moral baggage one would need on the road to progress, and a strong state would itself be the vehicle by which the trip would be made. To be sure, there were even then some who worried, not only about the highly materialistic underpinnings of the destination, but also about the bars on the windows—the extent to which individual life and liberty would have to be circumscribed. This bourgeois liberalism, even in the extreme form of *utilitarianism*,[9] unaccompanied as it was by any attempt at practical organization, actually made a greater impression in economics than in politics, where a new notion, modern *socialism*, now made its first appearance as a result of the Industrial Revolution.

The earliest proponents of socialism likewise did not make much of a showing, except by influencing later theorists. They were mainly utopians such as the Comte de Saint-Simon (1760–1825), who looked ahead to an egalitarian society based on an equitable distribution of the new abundance. Robert Owen, in *A New View of Society* (1813–1814), argued for a system of cooperatives that would do away with competition and exploitation, of the sort that only dedicated idealists such as the founders of the early Zionist kibbutzim in Palestine or Latter-day Saints settlers in Utah have proved able to make work even partially. Owen was at first merely a critic of the system that he had himself, as a successful cotton industrialist, helped to evolve. He assumed that all he had to do to have his views accepted at once by every reasonable man was to proclaim them. That approach failing, he appealed directly to workingmen and presently found himself, somewhat to his astonishment, the leader of the first trade union and cooperative movement—a role for which he was ill suited. Resolute opposition by employers and by the government suppressed the movement, which broke up in 1834. Another, equally disorganized movement was *Chartism* (1838–1848), whose program consisted largely of demands for parliamentary reform and for peasant homesteads in collective settlements.

Meanwhile, on the Continent, socialism was making slow progress during the turbulent period between the two revolutionary years of 1830 and 1848, highlighted by such episodes as the violent uprising of the Lyon weavers in 1831. But the political consequence of the Industrial Revolution that was of the greatest long-range significance was the publication of the *Manifesto of the Communist Party* in 1848. Where the Chartists had argued for more political power and the Socialists had plumped for the use of property for public welfare, the Communists boldly challenged the sanctity of private property altogether and promised to centralize all instruments of production in the hands of the state once the class struggle had resulted in the victory of the proletariat. With its impassioned call for a forcible overthrow of all existing conditions by concerted international action ("The proletarians have nothing to lose but their chains. . . . Working men of all countries, unite!"), the Manifesto was destined to dominate all left-wing theory and practice from then on. It was the work of two young German bourgeois, Friedrich Engels (1820–1895) and Karl Marx (1818–1883), who met in Paris after moderately successful careers, Marx as a journalist and Engels as a junior executive in a Manchester cotton factory of which his father was part owner. (Money provided by Engels *père* was to tide both men over hard times throughout their lifelong association.) The great significance of the Manifesto lies in its practical results, the formation of socialist and communist parties all over the world, whose organizations are thus palpably connected to the Industrial Revolution: the First International of 1864, the Second International of 1890 (from which present-day Labour, Social Democrat, and other socialist parties are descended), the Third International of 1919 (the ancestor of the Soviet, Chinese, and other Communist parties), and even the Fourth International founded in 1937 by the Bolshevik dissident Leon Trotsky (1879–1940).

TRANSFORMATION OF SOCIETY BY THE INDUSTRIAL REVOLUTION *extends beyond the technical to the economic, social, and political aspects.* First in Britain and then in an increasing number of other nations, industrialization led to rising living standards, urbanization, and a fantastic, sustained multiplication of goods and services. It also led to the decline of the importance of land in the economic scheme. One of the consequences was the abolition of slavery, an accomplishment toward which enlightenment, declarations of the rights of man, abolitionist sentiment, and all the rest contributed less than the existence of the industrial wage worker. It simply cost more to maintain slaves than to pay for cheap labor. National differences apart, it is a fact that emancipation proclamations antedating industrialization, such as those of Joseph II of Austria in 1781 and Alexander II of Russia in 1861, remained largely ineffective in improving the lot of the serfs; and slavery in the southern USA would doubtless have been abolished without war (as it was in Brazil ten years later) once the lure of escape to industrial jobs in the north made the guarding of slaves more costly than paying them wages—although the role of the Civil War in hastening industrialization must not be underestimated.[10]

Industrialization thus emerges as a pervasive agent of change, capable of turning an agricultural and commercial economy relying on simple tools into one characterized by a mechanized and urbanized factory system and (in a more mature embodiment) by a dominant capital-goods industry; of making irreversible changes in the social structure; and of providing the muscle for revolutionary political movements that would otherwise remain futile theoretical exercises. That such far-reaching transformations in both the public and private weal should take place without some considerable cost is not to be expected. That such costs should be transitory is to be hoped for. That most of them should fall on one sector of society is not to be borne.

The Crystal Palace, housing the first world fair, The Great Exhibition of the Works of Industry of All Nations in London in 1851, was itself a great engineering achievement featuring many innovations. It was completed in four months and stood (after being moved) until 1936, when it was destroyed by fire. (*Illustrated News,* 1851)

Chapter 2 Coming of Age of Technology

CONTINUING EVOLUTION OF INDUSTRIALIZATION, *the irreversible development of a society once it is successfully launched on the road to industrialization,* has been one of the outstanding characteristics of this process. In that sense, although one can clearly describe a society as "preindustrial," it is misleading to speak of a "postindustrial" society, which implies a steady state without further development. Nor may we ascribe literal meanings to a division of the world into "developed" and "developing" countries, which has become a sort of shorthand notation among experts, many of whom feel that to label a country tactlessly as "underdeveloped" has the double disadvantage of being both inexact and rude. After all, most present-day societies are still "developing." It is even inexact to refer to a specific industry—say, the electrical power industry—as "mature," simply because the basic development of most of its devices and processes was completed years ago. No sooner do we label such an industry as mature than along comes an innovation such as high-voltage transmission of direct currents or generation of power from nuclear sources to provide rejuvenation. What is historically unique about industrialization is, first, the sudden change in the rate of change, the way in which a sharp break shows up in the slopes of all the indices by which material progress over time can be measured, and second, the fact that these slopes have remained steep and the index curves have so far shown little inclination to revert to their near-horizontal preindustrial state—a state of affairs that obviously cannot continue indefinitely.

We have seen how the Industrial Revolution changed Britain from a small agricultural nation to the workshop of the world. At midcentury, while most of Europe was writhing in 1848 and all that, Britain was producing annual cotton cloth yardage by the billion, about half of the world's iron and most of its coal, and steam power equal to that of a million horses. Attempts to keep technical innovations at home were fruitless; they spread inexorably to the Continent and to North America and contributed to local development, especially in countries with abundant coal deposits or other sources of power. Britain continued to lead in a number of fields, and many important technical innovations were first made in British workshops, such as the Bessemer process of steel making, the construction of iron (and later steel) ships, and the development of quite elaborate machine tools (i.e., tools operated by machinery). But other countries, notably the USA, France, Belgium, Germany, and other regions in northern and central Europe, ultimately began to catch up, for the most following the British model.

In some fields Britain was presently overtaken. They were mainly the fields for which an advanced technical education was necessary, as in the chemical and later the electrical industries. For a variety of quite superficial reasons, ranging from academic snobbery to national prejudice against the excellent ideas of Queen Victoria's German-born consort Prince Albert, the system of university-level technical education that developed in England was rudimentary compared with that in other countries—a setback that was to plague the British well into the 20th century.[1] But as they prepared for the first world's fair, "The Great Exhibition of the Works of Industry of All Nations, 1851" organized in London under Prince Albert's aegis, and housed in an immense glasshouse of very original design dubbed the "Crystal Palace," the British had little inkling that their hard-won position as Top Dog was about to be challenged.

The transfer of technology between 19th-century Britain and other regions has recently attracted the attention of scholars, since the information thus unearthed may be useful in solving some of the similar problems of our day.[2] The subject need not occupy us here; rather, let us highlight some of the developments of the period that followed the Crystal Palace exhibition.

The principal technical aspects of a well-developed industrial economy are the ability to command ample sources of energy, to process materials, to enlarge transportation and communications, to engage in large-scale building and construction, and to improve methods of producing and distributing consumer goods, including food and other products of agriculture. It is obvious that advances in these fields have far-reaching consequences on the quality of human life, both good and bad, in terms of which the success or failure of technological advances may be measured. Before we can concern ourselves with evaluating such consequences, we need some acquaintance with the major technical innovations in each of the listed categories.

ENERGY CONVERSION *is the transformation of energy from one form to another.* The steam engine was the first effective device for energy conversion, from heat to mechanical energy. Earlier sources of energy—animal and human power, wind, and water—had merely served to translate one sort of mechanical power into another. To be sure, animal and human power is derived from food and thus ultimately from solar power, but the process is quite inefficient. What makes the use of fossil fuels more efficient is that burning coal or petroleum allows short-time use of solar energy stored over a long time. Even so, the earliest steam engines were most inefficient; in fact, there was no theoretical basis on which to express efficiency, which is the ratio of output to input. If the input is a coal fire and the output is a turning wheel, how are they to be compared? James Watt had no idea, beyond comparing the output of his engine with that of another run by a horse (hence, "horsepower" as a unit of rate of work, or power). Even the first attempt at optimizing engine design, made in 1772 by John Smeaton (1724–1792) on the Newcomen engine invented 60 years earlier, was largely a matter of trying out a few dozen samples and noting which combinations of part sizes worked best. (Smeaton, who was one of the outstanding engineers of all time, also pioneered in the design of canals and bridges and contributed to the improvement of windmills and water wheels.) The birth of thermodynamics, a towering scientific achievement, was a direct result of the interest created by the steam engine.[3] That is only one of the instances in which the development of a branch of science was preceded and stimulated by related technology. (Other examples are the effects of mining on metallurgy, of ballistics on mechanics, of "useful arts" such as dyeing on chemistry, and of radio-propagation studies on astronomy and planetary physics.) It is thus not surprising that the new science of thermodynamics found rapid application in the design and utilization of steam engines. The first major textbook, *Manual of the Steam Engine and Other Prime Movers,* by William John Macquorn Rankine (1820–1872), came out in 1859.[4] But not all who wanted to try their hand at improving steam engines were content to wait for scientific elucidation; nor did the development of competing devices stand still in the meantime. It was found, for instance, that new methods of feeding the steam to the cylinders and even using the same steam in several successive cylinders at different pressures and temperatures ("compounding") would improve efficiency. The water turbine, long known, was developed as a source of power. It was a new and better sort of wheel that turned by reaction to jets of water coming out at the edges. (A familiar type of rotating lawn sprinkler illustrates the principle.) This device was destined to have a useful life far beyond its use as a direct source of mechanical power. It proved to be the ancestor of the steam turbine, which ultimately displaced the steam engine, and of certain types of jet engines; and it found an important application in driving electrical generators.

Electricity made its technological debut as a source of lighting in the 1870s, after being known for centuries as a scientific phenomenon of little practical utility. Earlier, gas lighting had developed from its modest beginnings around 1800 into a big industry which changed city living, factory working conditions, school hours (the advent of adult education centered around evening study can be traced to effective lighting), and other aspects of the new society. This industry used coal gas—not natural gas from petroleum fields, which is a much more recent development—obtained at first by processing coal solely for this purpose but later increasingly as a byproduct in the manufacture of coke. Coal gas also provided a new fuel for cooking, heating, and industrial processes. Coal gas works better as a fuel than as light, since its normally bluish flame gives little light unless it is directed to play on a foreign substance such as lime (hence, "limelight") or the incandescent fabric "gas mantle" that Carl Auer von Welsbach (1858–1929) invented only in 1885, six years after Thomas Alva Edison (1847–1931) invented his incandescent electric bulb.

The Welsbach mantle prolonged the use of gas for illumination for some time after electric lights were introduced, but gaslight lost way to electricity steadily over the years and has almost completely gone out. Electricity in turn found uses other than lighting: in urban and interurban transport, in providing easily controlled motive power at sites far away from energy sources, and in numerous electrochemical processes. The source of mechanical energy converted into electricity was at first almost exclusively running water, and mountainous regions such as Switzerland, Italy, or California still depend a good deal on hydroelectric power. But eventually the steam plant became the largest source of electricity in most industrialized countries, with coal or oil as the source of heat energy to make the steam.

In the technically most advanced countries, steam plants derived from nuclear power are now coming to supplement the fossil-fuel plants, a trend of great importance to countries that lack fossil fuels but also to countries that have large reserves of such fuels but are concerned about adverse environmental effects of coal and oil plants. (To be sure, people are also concerned about the safety of nuclear plants.) In a steam plant, the heat energy of the fuel is converted into mechanical energy—the turning of a turbine—which in turn is converted into electrical energy by an electric generator coupled to the turbine. The next logical development would be to think of a way of eliminating the intermediate mechanical step and to achieve "direct" energy conversion from heat to electricity without mechanical motion. The feasibility of various methods has been demonstrated, but none has proved economical as yet.

MATERIALS PROCESSING *is the handling of materials and changing them for some specific technological purpose and use.* The 19th century saw great advances in the extraction of metal from ore: a metal such as nickel had been in a class with silver as a precious metal until then. Copper (and later aluminum) owes much to the electric industry, where it found many uses ranging from telegraph cables to electrical machinery and power distribution systems; this debt was repaid when electrolytic processes requiring large-scale consumption of electric power came into wide use in the reduction of both metals from the ores. Iron production was greatly facilitated and advanced by the substitution of coke for charcoal in smelting. But the greatest advance came in the production of steel, a high grade of heat-treated, hard iron containing carefully measured amounts of carbon and other elements. Steel had been so expensive that its use had been limited to a few implements such as springs, knives, and certain tools. Various makers of iron had discovered that blowing air through molten iron removed impurities by oxidation and at the same time produced enough heat to bring the liquid to the higher melting point of the pure metal. The first to devise a method to apply this technique to large-scale steel production was Henry (later Sir Henry) Bessemer (1813–1898). His method was immediately adopted in Sweden, Britain, and other countries—most importantly the USA, where the large distances between population centers created a railroad demand for steel that repeated the British experience with a vengeance. By 1890, the USA was the leading producer of steel, and has been ever since. Men such as Andrew Carnegie (1835–1919), the son of a jobless Scots weaver who had taken his family to Pennsylvania, became multimillionaires through their control of a good chunk of America's steel industry.

Up to that point, almost all technical advances in both ore reduction and the shaping of the metal products had been made by practical metallurgists with very sketchy scientific training. A man such as Bessemer certainly knew little chemistry.[5] After he discovered his process, he got his first encouragement not from a scientist but from an engineer, George Rennie (1791–1866), who told Bessemer to read a paper before the British Association for the Advancement of Science and to get some scientific counsel. But the generation of metallurgists that came immediately after Bessemer caught on very quickly to the value of science in studying the microscopic structure of metal samples, developing new alloys, and producing new materials at competitive prices. In fact, chemistry was the first science to interact systematically with large-scale technology, not only in the extractive industries and the utilization of their by-products, but also in the making of such organic materials as rubber and its substitutes, celluloid, and synthetic dyes. Other branches of technology were slower to turn to science for help. The wholesale deliberate application of scientific methods to technology and the quick derivation of entire industries from new scientific results are much more recent phenomena, typical of 20th-century technology—a characteristic that was thus prefigured by the interplay between chemistry and chemical engineering that began in the mid-19th century.

TRANSPORT, *the means of conveying people and goods from one place to another,* was limited to waterways and animals or animal-drawn vehicles until the advent of the railroad. That improvements were needed in regions that were in the process of industrialization was evident. To some extent the need was met by the building of canals between centers such as Liverpool and Manchester or Albany and Buffalo (the Erie Canal, completed in 1825), along which draft horses pulled flat-bottomed barges filled with bulk goods. In some countries, canal transport (now largely motorized) still remains competitive, but in most it has been replaced almost entirely by rail and road transport.

The interplay between railroad and industry, by which railroads not only served industry but also became prime customers for its products, was repeated by shipping. As steamships replaced sailing vessels, bold men undertook to use iron (and later steel) in constructing hulls and to add screw propellers at the stern to augment the paddlewheels on the sides. Beginning in the 1830s, ships of increasing tonnage began to come from British and American shipyards. The career of the great British engineer Isambard Kingdom Brunel (1806–1859) illustrates the trend: he built the wooden *Great Western* of 2300 tons in 1838, the iron *Great Britain* of 3000 tons in 1845, and the iron *Great Eastern* of 18,900 tons in 1858. The last was five times bigger than any other ship then operating. It also put down the first permanent transatlantic telegraph cable in 1865 and thus contributed to yet another branch of technology.[6]

The other great event in transport was the development of the internal combustion engine, prototype of the automobile, aircraft, and railroad engines of a later day. The groundwork was done well before the end of the 19th century. Motivation was provided by the patently low efficiency of even the best steam engines, in which heat was wasted all over the place. Various inventors had tried to make engines in which successive charges of illuminating gas or other substances would be made to explode to produce motion. If combustion could take place internally and without the wasteful step of producing steam, surely one could get more work out of an engine. Moreover, such engines would be potentially smaller and easier to start than large steam engines, making them available for running small shops and even individual machines. The first gas engine to achieve commercial success was designed in 1867 by Nicolaus August Otto (1832–1891) and was followed in 1876 by his four-cycle combustion engine fuelled by a gas-air mixture, about twice as efficient as the best steam engine of comparable size and less noisy. The "Silent Otto" was a great success and thousands were manufactured during the few years that remained before the ascendancy of the electric motor. But the Otto engines and competing devices were all stationary. If they were to be used to power vehicles, they would have to weigh much less, turn much faster, and use a liquid fuel. Such engines were devised for the first automobiles in Germany in the 1880s by Gottlieb Daimler (1834–1890) and Karl Benz (1844–1929), working at first independently but later joining forces in the firm that still bears their names. The first automobile engines weighed around 100 kg and developed about 1 hp; but by 1901, when the first Mercedes was introduced, its 200-kg engine produced 35 hp and speeds of more than 80 km/hr. The growth of this engine (and of its contemporaries in other lands) into today's automobile and aircraft industries, with their profound effects—good and bad—on contemporary life, is one of the most far-reaching achievements of technology.

Other relevant inventions were the use of tar, asphalt, and concrete in surfacing roads made of small broken stones (macadam) on a raised roadbed sloping to the sides, a configuration first proposed by John Loudon McAdam (1756–1836); and the vulcanization of rubber by Charles Goodyear (1800–1860) in 1839, which made commercial use of rubber (notably in automobile tires) possible.

One other invention has contributed enormously to road, rail, and ship transport: the diesel engine, a high-pressure internal-combustion engine burning cheap fuel at high efficiency and requiring no spark ignition, which Rudolf Diesel (1858–1913) patented in 1892. Although somewhat slower to develop than the low-pressure gasoline engine, the diesel engine ultimately came to predominate in the vast field that lies between the small gasoline engine and the huge steam turbine; today, it dominates all heavy land transport and most shipping.

The light internal-combustion engine made aviation possible. The first heavier-than-air craft in which Orville Wright (1871–1948) and Wilbur Wright (1867–1912) achieved controlled flight in North Carolina in 1903 was powered by such an engine; so were all other aircraft until the development of jet engines in the mid-20th century.

The internal-combustion engine also set off far-reaching changes in everyday life and in the very appearance of industrialized countries, ranging from such close dependents as the spare-parts industry and highway construction to the more removed motel industry, drive-in movies, and suburban living. But the greatest effect of improved transport technology has been the tremendous increase in the volume of commerce, the traffic in goods, which moves raw materials and finished products from country to country by the thousands of tons in a single shipment and contains the means of bringing the new abundance to all corners of the earth.

COMMUNICATIONS TECHNOLOGY *encompasses both the means by which messages pass between individuals and the methods of informing large groups.* Both were radically changed during the 19th century. Such advances as the introduction of the cheap, mass-produced graphite pencil and the steel pen nib; the drop in the price of newsprint and other paper; the marketing of inventions such as the rotary press with a cylindrical type-bed (1847), the type-writer (1874), the phonograph (1877), the linotype (1886), and the popular camera (1888); and the advent of motion pictures (1893) all contributed toward making communications easier and gave rise to huge industries whose descendants can be clearly seen today. Among other results of technological advances were the introduction of the one-cent newspaper in New York in 1833 and of the penny post in Britain in 1837; the reduction in cost was an innovation that soon spread to other industrialized countries.[7] But the most spectacular advances were made in the speed of communication. As a result of the rise of electrical technology, messages could now be sent with the speed of light.

The first attempts at an electric telegraph were either primitive bell-ringing devices or quite complex systems requiring several wires and "needles" to indicate letters of the alphabet. Such devices found limited use, mainly in railroad signaling. The real breakthrough came with the system developed by the American portrait painter Samuel Finley Breese Morse (1791–1872), which required only a single wire through which current was sent in short and longer bursts arranged in a code corresponding to the letters of the alphabet and numerals. Having successfully demonstrated transmission over moderate distances, Morse talked the U.S. Congress into funding a longer line, between Baltimore and Washington (about 50 km), in 1843. (Since the Morse code made no provision for punctuation, it is not clear whether the first message, "What hath God wrought," was an exclamation or a rhetorical question.) The telegraph succeeded very quickly, usually paralleling railroads and highways, but sometimes (as in the American West) striking out across the prairie to bind faraway territories to population centers. Western Union Telegraph Co. was organized in 1856 and linked the Pacific coast to the Atlantic. A submarine telegraph cable linked Britain with France in 1850 and with America, after several false starts, in 1865. Edison's first profitable invention, made when he was 23, was an instrument by which the telegraph was adapted for use on the stock exchange. Reuters began in 1851; General News Agency (predecessor of Associated Press) in 1856.[8] Almost from the first, the telegraph found its principal use not in conveying person-to-person messages but in commercial transactions, government traffic, and mass communications.[9]

That was less true of the telephone, invented by Alexander Graham Bell (1847–1922) in 1876. Nor did the telephone give the telegraph much competition at first, for technical reasons: before the development of amplifying "repeaters" located at suitable intervals, the weak telephone signal was rapidly attenuated over even moderate distances. Sixteen years elapsed before an erratic telephone circuit linked New York with Chicago. The telephone connection between the Atlantic and the Pacific coasts was not made for nearly 40 years—not until after triode (vacuum-tube) amplifiers became available—just in time for the 1915 San Francisco exhibition celebrating the opening of the Panama Canal.

The massive advances that followed the invention of radio ("wireless") telegraphy began around 1900, largely as a result of the technical and promotional activities of an Italian amateur scientist, Guglielmo Marconi (1874–1937). (Marconi worked in Britain, but Russian, German, French, and American contributions came almost simultaneously.)[10] In 1901, six years after his first experiments, he was able to bridge the North Atlantic and to communicate with ships far out at sea; but overland and between points connected by submarine cables, radiotelegraphy could not compete economically with wire telegraphy until many years later. The two systems coexist to this day, even after the introduction of many refinements (including telephone cables, microwave relays, and satellite repeaters), side by side with the telephone. Radiotelephony came into being at the beginning of the century; broadcasting did not begin until 1920, following technical advances made during the war of 1914–1918, and rapidly established itself as a means of mass communications. However, radio did not supplant newspapers and magazines; nor was it itself supplanted by its descendant, television. Communications technology is thus like many other human undertakings in that demand continues to grow and is stimulated by innovations that add to the sum of man's experience without necessarily doing away with what has gone before.

BUILDING TECHNOLOGY, *the application of advances in construction and architecture to buildings and other structures,* took a huge jump during the 19th century. Large-scale construction had been a feature of many earlier civilizations. The greatest of the Egyptian pyramids was reckoned as one of the Seven Wonders of the (ancient) World; all seven were construction projects. The Roman aqueducts, the Hagia Sophia, and the Great Wall of China were all completed long before industrialization was thought of.[11] Large public buildings, especially those intended for worship, were not exactly a novelty: the English cathedrals go back to the Middle Ages and the imperial mosque at Isfahan was completed in 1611, St. Peter's church in Rome in 1626, the Taj Mahal in Agra in 1648, St. Paul's cathedral in London in 1710. But each such structure was a *tour de force*, conceived by an outstanding architect and often decades in being erected. The need for the sudden profusion of factories, bridges, earthworks, city buildings, and structures of all kinds precipitated by the Industrial Revolution called for technical advances of a different order. The two most important were the use of new materials, notably iron and modern concrete, and the invention of new structural forms.

The use of iron frames in buildings was pioneered in Britain and France during the first half of the 19th century. In America, where timber was plentiful, wooden construction continued to predominate even in the largest structures, and much ingenuity went into the design of trusses and other arrangements by which the weaker material was made to do the work of iron. The largest domed structures of the new age, such as city markets and especially metropolitan railroad stations, which have been called the cathedrals of the industrial age, offered a particular challenge to the designer—a challenge that was met in different ways on the two sides of the Atlantic. While St. Pancras station in London was being covered by a single wrought-iron vault spanning over 70 m, an American structure of similar shape, the equally large Mormon tabernacle in Salt Lake City, got a wooden roof truss. Both buildings were still in use a century later, but the tabernacle roof had become something of a museum piece, the most monumental example of a method that was in the process of being replaced already at the time of its inception by iron structure even in America, as the supply of timber began to give out.

Iron frameworks made possible the development of the tall building, as epitomized by the Eiffel Tower erected for the Paris Exhibition of 1889 (the centenary of the French Revolution) by Alexandre Gustave Eiffel (1832–1923). For more elaborate high structures even iron proved to be too weak; the use of steel, first in iron-and-steel buildings and then increasingly in the form of all-steel frameworks, was the solution to the problem of erecting skyscrapers, the first of which were built at about the same time as the Eiffel Tower. (Their advent also had to await the development of a safe and successful electric elevator.) The basic idea of a steel framework and a "skin" of brickwork or masonry that has little structural purpose and serves mainly to keep out the weather and noise has been kept on in present building practice. The skin is more likely to be of concrete, aluminum, or glass nowadays, but even that is not such a new concept; the Crystal Palace of 1851 was an iron-frame building sheathed in glass, and the use of concrete walls (first introduced for fire protection) goes back to some of the earliest tall buildings.

Concrete had in fact enjoyed a boom since the discovery of artificial means for producing cement mixtures that hardened in a damp atmosphere, best of all under water. Natural mixtures with this property were known to the Romans, but the art of using such materials had been lost and was not rediscovered until 1756 by John Smeaton. The beginnings of a large-scale cement industry were not laid until well after the development of the synthetic material known as "Portland cement" in the 1850s; the first patent on it was granted to Joseph Aspdin (1779–1885) in 1824. The popularity of such a material for massive structures, especially those around water, as in harbors and bridges, did not extend immediately to buildings, since ordinary concrete cannot withstand tension (stretching) as well as it does compression. This problem was solved by the use of iron rods (or a whole mesh of them) put into place before the mixture was poured, so that the iron became embedded in the concrete and strengthened it, without itself becoming exposed to the weather. Such "reinforced concrete" has been the mainstay of large-scale construction ever since; the only significant new developments have come in methods of casting concrete, as in the construction of thin shells, and in the technique of stretching the reinforcing rods during casting, so as to counteract the stresses that the load would later apply ("prestressed concrete").

One other development deserves to be mentioned: the use of relatively thin boards closely spaced in the framework of domestic buildings, replacing the heavy timbers of an earlier day. This so-called "balloon frame" was invented in Chicago early in the 19th century, at about the same time that cheap nails became available, and remains the preferred method of erecting frame buildings to this day.

MASS PRODUCTION, *the manufacturing technique by which precision-made, interchangeable parts are rapidly assembled in a continuous, synchronized operation to yield a standardized unit,* is mainly a 20th-century development. Its antecedents go back to the Industrial Revolution. We catch a glimpse of one or another of its several characteristics even earlier, for instance in the use of bricks (interchangeable parts) or in a grist mill with chutes and conveyors (continuous operation). But the large-scale application of all the characteristics of mass production came with the rise of the American automobile industry, first in the Cadillac Motor Car Company of Henry Martyn Leland (1843–1932) and then, on a really massive scale in the Model T, the 1908 brainchild of Henry Ford (1863–1947). Output rose from 300,000 cars in 1914, when the first assembly line was introduced at Ford's Detroit plant, to 2 million in 1923; before the model was discontinued in 1927, 15 million had been sold. The Model T also showed that mass production does not necessarily result in inferior products of little merit; rather, it is a means of manufacturing quite complex items at a low unit cost. The price of the Model T had dropped to $290 by 1924, so that the factory workers themselves could afford it—for in 1914 Ford had created a sensation by introducing the 8-hour day and the unprecedented high wage of $5 a day, supplemented by a profit-sharing plan under which up to $30 million was distributed to the workers each year.[12] (Ford's paternalistic attitudes and antediluvian views on other social and political questions made trade-union leaders hate him, but it is a fact that the United Automobile Workers were unable to unionize his plants until 1941, long after those of his competitors.)

Mass production is thus linked to a mass market, either potential or—in more recent times—one that can be created largely by advertising and other means. The principle that workers engaged in mass production, suitably rewarded, also become consumers of their own products (including housing) played an important part in making the American standard of living the highest in the world—certainly a greater part than, say, social legislation, a field in which until recently the USA has remained decades behind many poorer countries. (Social security was introduced in Germany in 1883;[13] in the USA, in 1935). Moreover, the American trade-union movement, led by Samuel Gompers (1850–1924), from the first stressed economic participation (the "piece of pie" philosophy) over ideological preoccupations. The lack of class consciousness in the highly mobile society of the USA doubtless nurtured (and was in turn fed by) the worker-consumer conjunction, a scheme that is only now catching on in other countries in which rising levels of prosperity have begun to blur class distinctions. It is the capitalist's answer to the socialist: that private ownership and enterprise, *when coupled with widespread participation in the fruits of increased productivity,* are likely to lead to real improvements in the well-being of the largest number and to a greater social justice more quickly than the choicelessness and lack of incentives characteristic of public ownership; and that social justice is next to meaningless in hopeless poverty and under political coercion. But industrialization is greatly hampered in those nonsocialist countries whose inegalitarian social systems hinder widespread economic participation, for example, in most Muslim and Latin American lands; for those people in such countries to whom outside information is accessible, the socialist alternative must loom very attractive indeed.

Mass production quickly spread beyond the automotive industry and now predominates in the manufacture of all goods in industrialized countries. It was accompanied by other rationalizations: of *management,* notably through the efforts of Frederick Winslow Taylor (1856–1915) and his successors, whose time-motion studies ("Scientific Management") laid the foundations for an important part of industrial engineering;[14] of *distribution,* through such innovations as department stores, chain stores, self-service, mail-order houses, mass warehousing, and (more recently) supermarkets and shopping centers; of *marketing,* the systematic study of consumer needs and wants (and often their stimulation through advertising and extension of consumer credit) and the planning of production so as to meet them; and of *industrial research,* the systematization of invention by which individuals and teams working in industrial laboratories, often on prescribed tasks, have largely replaced (but not entirely eclipsed) the lone inventor laboring privately under adverse circumstances toward wealth and movie fame.

Mass production also created the specter of dehumanization, the psychological (and often physiological) hazards of the sort of mindless, repetitious work epitomized by Charlie Chaplin's famous assembly-line worker in his movie *Modern Times* (1937), whose sole task is to tighten a pair of nuts as unit after unit speeds by on a belt moving at a rate over which he has no control. High pay and even job security do not compensate the worker for the feelings of anonymity and uninvolvement that go with a deadening, depersonalized task in which achievement cannot be measured, except with a stopwatch. The role of such factors in motivating workers was first clearly brought out during the 1920s in the famous "Hawthorne experiment" carried out by Elton Mayo (1880–1949) and his associates at the Western Electric Company's Hawthorne Works near Chicago.[15] The objective was to find out which factors—lighting, ventilation, humidity, length and frequency of rest periods, temperature—affected the time needed to assemble a simple part of a telephone switch. To their astonishment, the researchers found that almost any change produced an improvement, including changes from better lighting to poorer, and from a condition whose introduction had resulted in an improvement back to the original condition. The greatest improvements came when the workers were asked for their opinions before changes were made and were allowed to set their own pace. The conclusion of this pioneering study was that the sense of participation and involvement that came from the simple fact of being under investigation was the all-important factor.

The Hawthorne experiment marked the entry of the social and behavioral sciences into industrial management, in which they have since become well established. They contributed to the growing realization that human effectiveness cannot be measured merely by rising efficiency; and that the measures that are most likely to lead to involvement, to active rather than merely passive participation in the industrial process, are those that are compatible with such age-old humanist concerns as self-respect and human dignity.[16] Trends toward too much specialization of the assigned work task have been arrested and in some respects reversed, as in the increasing practice of rotating teams of workers from job to job (with apportionment of tasks within the group left to the team itself); of "job enlargement," a method by which a simple task is increased in scope (for example, by including setting-up and checking procedures previously carried out by others) and the worker's skill—and involvement—are increased at the same time; and above all, of designing automated machinery to do some of the most boring jobs, an example of the way in which a problem created by technology is ultimately solved by more technology.

Techniques of mass production have also been applied to the growing, harvesting, processing, and packaging of food, in fields ranging from mechanization and partial automation of food production to novel methods of preserving (for instance, dehydration and freezing) and dispensing food. Modern fertilizers and control of pests have also played important parts. One of the hallmarks of industrialization is the low percentage of the population engaged in producing a country's food. Where development has been negligible, almost everyone works on the land. Where industrialization is well under way, for instance in the USSR, only half the population is engaged in agriculture. In a highly industrialized country, such as the USA, fewer than 10 percent work on the land (to be sure, supported by the chemical, agricultural machinery, and other industries) and produce surpluses as well.

HIGHER TECHNICAL EDUCATION, *the instruction in engineering and related applied sciences on the university level,* received its first impetus from the Industrial Revolution. Formal education was surprisingly slow to get going in England, where specialists continued to be trained by the system of pupilage well into the 20th century. Under this system, the "student" is apprenticed to a practicing engineer or engineering firm and is finally certified as competent by a nonuniversity professional body. (On the other hand, such bodies were established quite early in Britain: in 1818 for civil engineering, in 1847 for mechanical, and 1871 for electrical.) "Civil" engineer is a term by which John Smeaton described himself in the 18th century so that others could tell him apart from a military engineer. The first professional engineering schools were French: École Nationale des Ponts et Chaussées (bridges and roads, 1747); École Nationale Supérieure des Mines (mining, 1778); École d'Arts et Métiers (arts and crafts, 1788); and École Polytechnique (1794). The last remains a top school of engineering and science whose graduates man an astonishing number of top administrative posts in French industry, government, and the technical branches of the armed forces to this day. But it has also produced many fine scientists and supplied teachers for other institutions the world over. The first engineering professor in America, Claude Crozet (1790–1864) at the U.S. Military Academy in West Point, N.Y., was a graduate of École Polytechnique; his biographical entry in the officers' gazette notes that "in giving instruction to his pupils he made much use of the *black board* and taught Descriptive Geometry as a necessary preliminary to the proper study of Engineering," then a startlingly novel (and very French) approach to the subject.[17]

In Germany and Austria, existing technical secondary and trade schools began to develop into polytechnics early in the 19th century —the one at Karlsruhe was created in 1825—and ultimately achieved full professional standing and university rank as *Technische Hochschulen.* (Although the term translates as "technological high schools," a closer equivalent is "engineering colleges.") But university science departments also supplied industry with personnel, especially German chemistry departments, where postgraduate research assistants received training of great practical usefulness.

The first American engineering college (apart from West Point) was the Rensselaer Polytechnic Institute in Troy, N.Y., founded in 1824 on the École Polytechnique model. It became a prototype for other colleges of the same type, not a few evolving from trade schools such as Cooper Union (1859) and Pratt Institute (1887) in New York City and Throop Polytechnic Institute (1891) in Pasadena, California, which became one of America's top science schools after it was reorganized as California Institute of Technology in 1920. But the greatest boon to technical education in the USA was passage of the Morrill Land-Grant College Act in 1862, under which federal lands were set aside for colleges that were intended to supplement the instruction offered at such ancient seats of learning as Harvard (1636) and Yale (1701) by concentrating on "such branches of learning as are related to agriculture and the mechanic arts." They did that, and much else besides, for out of the more than 70 colleges that resulted grew some of America's greatest universities, private and public, coast to coast, from the Massachusetts Institute of Technology (1865) to the University of California (1868).

Harvard had added engineering to its curriculum in 1847 and Yale in 1852, but they (and other private colleges) could not have managed to supply all the talent needed by an industrialized country; nor could Oxford and Cambridge do the job for England. Almost everywhere else higher technical education was publicly supported by central, provincial, or municipal governments from the outset and engineering had full professional and high social standing. In Czarist Russia, for instance, despite the glamor of the Court, a civilian engineering career was valued well above a military career and on a par with one in medicine or the law. But in the USA (even after passage of the Morrill Act) and in England, engineering has long been a career by which the son of a workingman or an immigrant could advance himself to professional standing with minimum time expenditure: three or four years of full-time study were enough to reach the first professional degree, compared to six or more for law or medicine, and opportunities for part-time study were abundant. Until quite recently, most American engineering students were sons of blue-collar workers and somewhat outside the pale: at Yale, for instance, they were not accepted into Yale College fraternities until 1945, and the letter "S" (for Sheffield Scientific

School) and later "e" (for engineering) was added after the year designating the graduating class until 1966. In England (but not in Scotland) such attitudes were even more widespread, much to the country's detriment.[18] By contrast, in most countries whose systems of higher technical education resemble the German, including the socialist countries, the title abbreviation "Ing." or its equivalent precedes the bearer's name and carries a weight comparable with that of the "Dr." of a physician, lawyer, or other university graduate. (Incidentally, nearly half the engineering students in the USSR are women, compared with about 1 percent in the USA.)

The great increase of engineering college graduates has affected not only the technological sector of industrial society but also its entire structure. They are increasingly found not only in top industrial posts (where the owner's son, the businessman, and the accountant used to predominate) but also in government, politics, and other endeavors to which an engineering education might be considered as merely incidental. The view that such an education is narrow and unintellectual derives from *démodé* concepts of what it used to contain. The modern professional curriculum of a major American engineering college, for instance, is firmly based on such a rich selection of sciences, social sciences, and humanities that it might be justly considered as providing a more "liberal" (that is, broader) education than most curricula in the liberal arts.

Lastly, the importance of engineering to industrial society in countries where the engineering college is often part of a general university has very likely affected such universities themselves, not only through the presence of an additional (and identifiable) body of undergraduates with strong professional orientation and motivation, but also through the activities of engineering professors, which extend beyond technical instruction to participation in university administration and to even more esoteric tasks, such as the writing of general books about technology.

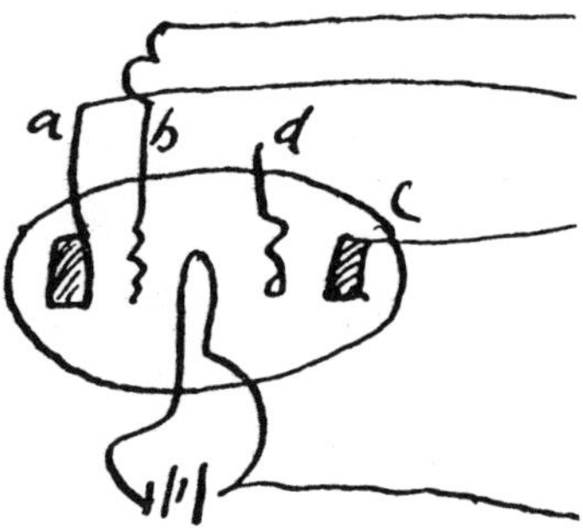

The triode patented by Lee de Forest in 1906 as "Device for amplifying feeble electrical currents" became the foundation stone of the electronics industry. Inventor's sketch shows the principal feature, the "grid" electrode (wavy line), whose voltage controls electron flow between other electrodes; small variations in grid voltage can be amplified into large variations of voltage in the external circuit. (Smithsonian Institution)

Chapter 3 Rise of Modern Technology

THE SECOND INDUSTRIAL REVOLUTION, *the recent qualitative changes in the relation of technology to science and in technology's aims*, is above all associated with the astounding developments in electronics that began to come to fruition in the mid-20th century. Science and technology have leapfrogged past one another throughout recorded history; now one was in the lead, now the other. As mentioned previously, anyone seeking a causal relation was just as likely to find technology the cause and science the effect as the other way around: gunnery led to ballistics, the steam engine to thermodynamics, powered flight to aerodynamics. But with the Second Industrial Revolution, the casual relation between technology and science changed to a systematic exploitation of scientific research—sometimes research undertaken with technical uses in mind. As we saw, in at least one sector of technology—the chemical and pharmaceutical industry—this change actually came a good while sooner, prefiguring the later experiences of the other "scientific" industries by fifty years or more. But there is a second characteristic of the Second Industrial Revolution that puts it squarely in the middle of the 20th century. Whereas the first Industrial Revolution saw machines put in place of animal and human muscles, the

Second Industrial Revolution began the use of machinery for some of the functions performed by the senses and the human mind.

The change from invention to industrial research is exemplified by the difference between Bessemer and his successors in chemical metallurgy (chapter 2). Another instance is the birth of the plastics industry. The first artificial plastic was celluloid, invented by John Wesley Hyatt (1837–1920) in 1869. Hyatt was a self-taught American printer whose principal aim seems to have been finding a cheap substitute for ivory billiard balls. He did employ a chemist, Frank Vanderpoel, and he did start a successful business, but for a long time the use of celluloid remained limited to small consumer products such as dental plates and detachable collars for men's shirts. By contrast, the next major contributor to plastics technology, Leo Hendrik Baekeland (1865–1944), got a first-class university education in chemistry in his native Belgium before he emigrated to the USA in 1889 and set about improving photographic paper and developing bakelite (in 1909), the synthetic resin whose widespread use in industry signalled the birth of the modern plastics industry. (Among its other major milestones: the development of rayon in 1924, of nylon in 1938, of polystyrene in the 1930s, of polyethylene in the 1940s, and of high-temperature plastics in the 1960s.)

The influence of scientific research is even more evident in the rise of the dyestuffs industry. Dyeing had depended from antiquity entirely on natural dyes, mineral or vegetable—such as ocher and indigo—until 1856, when William Henry (later Sir William) Perkin (1838–1907), an 18-year-old assistant at the Royal College of Chemistry in London, discovered the first aniline (coal-tar) dye, mauve. The first synthesis of a natural organic dye, indigo, following its analysis in 1883 by Adolf von Baeyer (1835–1917) at the University of Munich, was the result of years of painstaking laboratory work. Extensive research also preceded the introduction of synthetic drugs such as aspirin (1899) by the firm founded by von Baeyer's near-namesake Friedrich Bayer (1825–1880)[1] and salvarsan (1909) by Paul Ehrlich (1854–1915). Salvarsan, the "magic bullet" against syphilis, was at first called "606" because Ehrlich had systematically tried 605 other substances before hitting upon it. The same procedure had been followed by Edison in selecting a suitable material for light-bulb filaments: he succeeded after testing over 1600 incandescent substances in his industrial research laboratory at Menlo Park, N.J.

Edison's laboratory was the predecessor of the General Electric Co.'s Research Laboratory founded in 1901 in Schenectady, N.Y., one of whose employees, Irving Langmuir (1881–1957), was the first industrial research scientist, unconnected with any university, to receive the Nobel prize (1932), for his contributions to the chemistry of surfaces. (Adolf von Baeyer and Ehrlich also got Nobel prizes.) The idea of a laboratory through which industrial achievement would be closely tied to scientific research thus dates back to the 19th century. It has been greatly elaborated during the 20th century, many of whose most spectacular technological triumphs can be traced directly to the laboratory.

Coupled with the idea of industrial research is another idea—the notion that invention need not be invariably a flash of individual inspiration but can often proceed by a systematic analysis of possibilities and elimination of false trails toward a preconceived goal. The saying attributed to Edison comes to mind: "Genius is one per cent inspiration and ninety-nine per cent perspiration." But modern industrial research also seeks to cut the effort that is needed by consciously bringing every bit of knowledge and every method known to science to bear on the development of new materials, processes, devices, and systems to satisfy previously established requirements. This approach does not always yield quick answers—we are still a long way from a cancer cure or a cheap method of reducing car pollution—and we surely cannot do without inspiration. But much as we may regret it nostalgically, many technical innovations now incontestably result from systematic effort, often by teams of trained research workers, rather than from flashes of individual genius.[2]

This professionalization of technology, the handing over of functions carried out largely by artisans during the first Industrial Revolution to specialists with an increasingly scientific education, together with the growing dependence on science, is one dominant aspect of the Second Industrial Revolution. The other is man-made devices doing the work of human intelligence, largely the result of progress in electronics, which we review briefly before we discuss its achievements.

CONTRIBUTIONS OF ELECTRONICS TO THE SECOND INDUSTRIAL REVOLUTION *comprise both "hardware" (devices, ranging from transistors to computers) and "software" (methods of using hardware, such as communication theory and computer programming).* Electronics found its first industrial employment in radiotelegraphy (1896), which duly evolved into radiotelephony

(1906), radio broadcasting (1920), and television (1935), followed by voice communications via underwater cables (1956) and satellite repeaters (1962). The most important device of the first fifty years was the vacuum-tube triode amplifier of Lee de Forest (1873–1961), invented in 1906 but not in general use until 1913, when it was put to work in the repeaters of the first American transcontinental telephone network. The triode was the first device by which an electronic signal could be scaled up (amplified). The scaling could be made arbitrarily large, but at a certain point the circuit would suddenly burst into oscillations—adjusting some early radios was largely a matter of keeping them from producing ear-piercing whistles—a property that was soon put to use in a different circuit, the triode oscillator. The early radio amplifier that turned so easily into a generator of oscillations, the "superregenerative" receiver, was replaced by the whistle-free "superheterodyne" receiver and still later by the FM receiver employing frequency modulation, a short-wave method that is singularly free from distortion and interference. Rather remarkably, all three were developed by the same man, probably the greatest radio engineer of all time, Edwin Howard Armstrong (1890–1954), an engineering professor at Columbia University in New York City.

The uses of the triode amplifier and its companions—the light-sensitive phototube, the cathode-ray oscillograph, and other vacuum tubes of even more complex construction and function—soon went beyond telecommunications. Wherever an electronic signal needed to be generated, amplified, or otherwise modified, a vacuum tube stood ready. Electronic phonograph (gramophone) amplifiers played back music with high fidelity, the first step on the long road that leads from the crude mechanical amplifiers of Edison's day to the modern "hi fi" set. The recording of sound on film, the method by which sound motion pictures (likewise based on Edison's work) are made, utilized electronic devices from the first; so did the first electronic music instruments (1920), forerunners of a chamber of horrors extending from the electronic (that is, pipeless) organ to *musique concrète*, an art form in which recorded natural sounds are electronically manipulated and transformed, seldom to advantage. Sound amplification made a host of devices possible, such as public-address systems and hearing aids. The invention of the electron microscope (1930), a vacuum-tube device based on the cathode-ray tube and capable of much stronger magnification than the light microscope, brought electronics into metallurgy and biology. Other sciences also benefitted as electronic instrumentation took over in field after field. Even medicine began to take advantage of the new tool, for instance, in healing by diathermy, the heating of deep-seated tissues by high-frequency radiation. All these extensions of electronics beyond communications to the fields of entertainment and science are more or less familiar. But they are very new. As recently as 1940, telecommunications, electroacoustics, and scientific instrumentation occupied the attention of all but a tiny portion of the electronics industry. Electronic control of industrial processes was an esoteric specialty; electronic computers were unheard of; and radar was a closely guarded military secret.

INDUSTRIAL ELECTRONICS, *the application of electronic technology to the control of industrial processes,* got its start in the conversion of alternating current to direct current. Most electric power is generated in ac form but many devices require dc—for instance, the triode amplifier which, in early versions, had to be connected to three dc batteries, one of them of the wet accumulator type used in cars. These batteries, which had to be frequently taken in to be recharged, were replaced in the late 1920s by another vacuum-tube circuit, the rectifier, which performed the ac-to-dc conversion and made the first all-electronic home radio receiver possible, one that could be simply plugged into a wall socket. Other uses for the rectifier were soon found in industry, especially after variants were invented by which the magnitude of the dc voltage could be controlled. Now it became possible to vary the speed of motors, to regulate the progress of electrochemical reactions, to start and stop manufacturing processes—all electronically. Before long ways were found to do all these things automatically, according to a preset schedule. Moreover, electronic methods of holding a normally variable quantity at a steady value—say, the temperature of a furnace—became commonplace.

The technique that underlies all such control is the electronic version of the *feedback* principle, which we have already encountered in the mechanical form of Watt's fly-ball governor. Other familiar examples abound, not all drawn from technology. The thermostat that turns the central heating of a house on and off is actuated by a drop in temperature and interrupts the heating process when the preset level has been restored. Economic supply and demand is another example. Most systems operate better with feedback than without it, especially in the face of unexpected changes. If the thermostat is replaced by a timed on-off switch, the system performs much less satisfactorily if a sudden heat or cold wave strikes or the quality of the fuel changes; and a rigidly fixed economy, with wages and prices unrelated to productivity, cannot respond to a catastrophic change in demand or in the availability of critical raw materials nearly as well as a system geared to the market.

In an electronic amplifier, a small input signal is normally turned into a large output signal. If the input level varies, so does the output level. If we channel off a bit of the output signal and lead it back to the input, we increase the input, and hence the output, and hence in turn input and output, and so on, until the signal grows uncontrollably in a way familiar to anyone who has ever heard what happens when the microphone (input) of a public-address system is inadvertently turned toward the loudspeaker and so picks up its own output. That is positive feedback, leading to instability and unwanted oscillations. But if we interchange the wires so that the part of the output led back to the input is of the opposite sign and subtracts from the input rather than adding to it, we achieve control. An unwanted variation in input—for instance, the fading in and out of a long-distance radio signal—is counteracted by this negative feedback from the output, which is thereby held steady. This technique, as applied in the abovementioned automatic volume control (AVC) circuit of radio receivers, has been vastly elaborated and is used in countless variations to control not only volume (that is, magnitude) but also frequency (the AFC circuit that prevents drifting in the better FM receivers), temperature, size, quantity, and any number of other parameters in industrial applications.

One of the better-known elaborations is the *servomechanism*, a means of electronically controlling a machine by measuring the difference between a desired state and the actual state; this difference or error is then used as the actuating input signal that causes the machine to be operated in such a way as to reduce the error to zero. The servomechanism imitates physiological response. The action of picking up a glass, accelerating it and then decelerating it to the lips, tilting it to drink, and then setting it down again is carried out in response to a continuing stream of stimuli or clues, not merely visual (a blind person can learn to do it), which compare the position of the glass with the ultimately desired position and then command the muscles to reduce the difference to zero, without spilling, knocking out one's teeth, or breaking the glass by too tight a grip or by setting it down too hard, regardless of whether it is a delicate wine glass or a heavy tumbler. Similarly, a servomechanism can be designed which actuates the engines and control surfaces of an aircraft in such a way as to maintain a prescribed height, direction, and speed. Such an "automatic pilot" continually monitors the aircraft's actual height, direction, and speed, measures any differences between them and the preset values, and uses these differences or errors to initiate the corrective action to reduce them to zero. The servomechanism is thus an error-sensing electronic feedback circuit that holds the performance of a mechanical device to a prescribed objective. As in most feedback circuits, the power to operate the device is not supplied by the servomechanism itself; rather, it controls the amount of power supplied, sometimes merely by switching it on and off. A familiar example is the power-steering mechanism on some cars: a slight turn of the steering wheel cuts in a hydraulic mechanism that aids the steering motion as long as the pressure continues and then switches itself off again.

By a further refinement, it is possible to make a machine follow a complicated prescription. For instance, in the operation of a lathe, a rod of material to be machined turns on its axis while a cutting tool moves alongside; the machinist moves it in and out according to whether the cut is to be deep or shallow. The machinist's motions can be recorded on tape and identical additional pieces can be made by introducing a few motors and switches that let the lathe follow instructions from the tape rather than from the machinist. (A means of monitoring the depth of the cut and feedback of this information must also be provided, corresponding to the machinist's practice of checking the dimensions from time to time.) The replacement of an operator by automatic means of issuing and carrying out of instructions is a simple form of *automation*. In a more complex embodiment, the automated system also makes decisions, as in an automatic elevator, which ignores calls that would send it in the other direction until it has completed its travel in the direction it is going, but stores the information and uses it to make the requested stops on the return journey, moreover remaining stationary at a floor with its door open until all passengers have passed through them.

In a still more complex task, for instance the placing of a space vehicle into a desired trajectory, feedback also plays an all-important part. Data giving the vehicle's position, velocity, acceleration, and attitude are continually sent back to the control station, where they are compared with the values corresponding to the ideal trajectory; instructions are then sent back to correct any errors. But the mass of data pouring in and the near-instantaneous reactions needed make a very rapid method of dealing with data necessary. Such *data processing* is carried out by a computer, which has become an indispensable part of many control systems. The theoretical and practical developments that accompanied the birth of the electronic computer and its role in automatic control are described in the next chapter. The most important of these developments was the discovery of the transistor in 1947, which was followed by a whole family of characteristically small and low-power devices that have replaced vacuum tubes, mechanical switches, and other parts and have completely revolutionized electronics. Before we come to computers, it remains to describe radar. That development just preceded this revolution and is at once the last, greatest achievement of the vacuum-tube era and a spectacularly successful extension of human sense by electronic means.

RADAR *(acronym for radio detection and ranging) is a method of detecting obstacles in the path of a narrow radio beam and noting their direction and distance by measurements made on the echoes.* It was developed almost simultaneously in at least six countries in the mid-1930s, following the independent development of the necessary techniques (very short waves, reliable picture tubes, radio direction finding, and appropriate circuits) for other purposes.[3] The new method was developed most quickly in Great Britain, the country that not only was most concerned about enemy air attack but also had the best schemes for the design and operation of radar, following the ideas of Robert Andrew Watson Watt (later Sir Robert Watson-Watt). When war did come (1939–1945), British radar was ready. It has been widely credited with tipping the balance against the German attackers in the Battle of Britain in 1940, when data locating the attacking formations enabled the defenders to deploy their dwindling reserves to maximum effect.

The British radar of 1940, for all its tactical importance, was a relatively primitive affair in which a quivering squiggle on the face of a small, uneven picture tube had to be interpreted as to distance and direction from instant to instant. The more familiar version in which multiple targets are accurately and continuously mapped on a large screen is a later elaboration, but not much later. Most of the details had been at least thought of—and not a few had been actually realized—before the end of the war. The real innovation was the startling extension of the human senses by the use of a new engineering technique. To be sure, electronics had been used to extend human senses before. Speaking over a radio link with someone half-way around the globe or hearing or seeing a faraway event by radio or television are also extensions of our unaided sensory capabilities; so, for that matter, are the telescope, microscope, thermometer, X-ray camera, and hearing aid. But the dramatic debut of a method for detecting moving targets hundreds of kilometers away, through fog or in darkness, and pinpointing them on a map was an accomplishment of a different order. Since then, electronic means of replacing touch, smell, and even taste have been employed and electronic sensors far more sensitive and more quickly responsive than the human senses measure pressure, acceleration, pitch, temperature, color, and dozens of other parameters, usually for control purposes.

Radar itself has given rise to numerous offspring since its bow in air defense, not only in the adjoining areas of air and water transport, space travel, and space communications, but also in such seemingly unrelated fields as chemical spectroscopy, optics, atomic physics, and the industries derived from them. The Nobel prize-winning discoveries of nuclear resonance (Felix Bloch and Edward Mills Purcell, physics, 1962) and of the maser and laser (Nikolai Gennadevich Basov, Aleksandr Mikhaylovich Prokhorov, and Charles Hard Townes, physics, 1964) were directly related to earlier radar research by the recipients. Advances in the design of instruments for nuclear technology also owe a great debt to radar. Finally, radar gave a major push to the development of cybernetics, as we shall see in the next chapter.

"Microminiaturized" electronic circuit is traced from mask (shown slightly reduced, *left*) by a highly focused electron beam on a specially prepared surface, which becomes part of the integrated circuit (actual size, *right*) after further processing. (Hitachi, Ltd.)

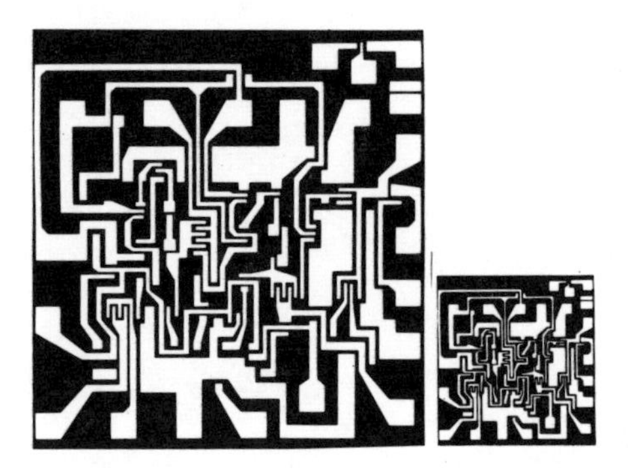

Chapter 4 The Computer Technology

CYBERNETICS *has been variously defined as the study of control systems comprising the brain, the nervous system, and electromechanical communication devices; and as the study of a much wider field.* The word was coined in 1947 by the American mathematician Norbert Wiener (1894–1964) from the Greek χυβερνήτης for *steersman,* from which the Latin *governor* is also derived. He developed a grand conception of the subject that included not merely engineering systems and physiological phenomena related to automatic control, but also "a new interpretation of man, of man's knowledge of the universe, and of society." These phrases, which are cited from the second volume of his autobiography, show how Wiener was led to view sociology, anthropology, economics, and even existentialist philosophy as branches of cybernetics.[1]

In the present chapter, we shall follow the less grandiose concept suggested by J. R. Pierce,[2] who considers cybernetics to include communication (or information) theory; smoothing, filtering, detection, and prediction theory; negative-feedback and servomechanism theory; and automata and complicated machines, including the design and programming of digital computers. Inclusion of even this more restricted list of topics under the designation *cybernetics* does not mean that they derive from a single, all-embracing concept that sprung ready-made from Wiener's Jovian brow. Most can be traced to entirely different origins, from which they grew vigorously and independently; their inclusion in cybernetics is a latter-day development. Recent developments suggest that some of these fields, for instance, the theory of automata, are acquiring the status of separate sciences whose practitioners consider being lumped under "cybernetics" as old-fashioned. However, the term has acquired enough popular currency to make its use convenient in this context, especially since the practical realization of the engineering consequences relies on much the same devices throughout. Foremost among them are the results of the developments that followed the invention of the transistor (a name coined by Pierce) in 1947, which we shall describe first.

SOLID-STATE ELECTRONICS *deals with electronic devices made of solids.* Such devices have been associated with electrical engineering from the first. All materials more or less conduct electricity: some, called conductors (such as most metals), do it particularly well; others, called insulators (such as glass and ceramics), scarcely at all. In between there is a class of materials called semiconductors.

The amount of electric current that passes through a conductor is directly proportional to the applied voltage (Ohm's law), but there are some materials or combinations of materials, mostly junctions of metals and semiconductors, in which the current flows much more easily from one end or terminal to the other than in the opposite direction. Such materials alternately act as conductors and insulators, depending on which of its two terminals is positive and which negative. If the direction of the current is continually alternating (ac), the result is that what gets through is only the positive half and the negative half is stopped; in other words, the current that gets through is all in the same direction (dc).

This capability of turning ac into dc, called "rectification," is also useful in radio, where it is called "detection."[3] A junction of a fine silver wire and a crystal of lead sulfide (galena) was widely used as a detector, known as the "cat's whisker," in the early days of radio, until it was replaced by the first vacuum tube, the diode (a two-electrode predecessor of the triode), originally proposed by Sir John Ambrose Fleming (1849–1945) in 1904. But during the war of 1939–1945, it was found that the Fleming diode did not work very well in the ultra–high-frequency range used for radar and the semiconductor diode made a comeback. The fields of solid-state physics and of engineering manufacture had both advanced considerably in the meantime and the "crystal diodes" of the 1940s were vastly superior to the cat's whiskers of the 1920s. Physicists were fascinated with the phenomena taking place at the junction of the two materials, where the point of the wire made contact with the semiconductor surface. After the war, a research group at Bell Laboratories in New Jersey undertook to study this critical region by probing the junction with a second wire, and then with a third, and measuring the currents flowing between them. This work led to the discovery that the size of the current between wires 1 and 2 depended on the voltage applied between wires 1 and 3. In other words, any variations in the voltage between a pair of wires were accurately reflected in variations in the current between another pair and could be easily transformed into voltage variations many times greater than the original ones. This configuration, dubbed the *point-contact transistor,* could be thus used as an amplifier (and hence also as an oscillator), replacing the triode vacuum tube. Unlike tubes, transistors required no vacuum, no heating elements, no warm-up time, and no high voltages; and they were smaller than the smallest tubes. For their parts in the discovery of the transistor effect, the three group leaders of the Bell Laboratories team, John Bardeen, Walter Houser Brattain, and William Shockley, received the 1956 Nobel prize in physics.

The transistors in general use today depend on junctions between slightly dissimilar materials and bear little resemblance to the original point-contact transistor. Automatic manufacturing techniques have been developed involving the production of millions of units from materials whose exact composition is controlled with incredible precision. The replacement of tubes by transistors, a process known by the horrific name of "transistorization," did not proceed very smoothly in the beginning. Low-power devices in which space was at a premium, such as hearing aids, were among the first to be converted. The progress of the hearing aid from a bulky tube amplifier (with heavy batteries and connecting wires between microphone, set, and earpiece) to an unobtrusive device that fits, complete with battery, into the tip of an eyeglass frame illustrates a process (known by the even more horrific name of "miniaturization") that also had a profound effect on the early stages of the space race. Massive concentration of research on rocket-propulsion fuels and techniques enabled the USSR to orbit the first artificial satellites, Sputnik I and II (1957), which made a few preliminary atmospheric measurements. The USA, whose rocketry research had been allowed to fall behind, would not have caught up for years if it had not been for the fact that its more advanced electronic technology enabled it to cram much more instrumentation into a lighter satellite. As it was, the very first American satellite, Explorer I, was launched less than four months after Sputnik I and was responsible for a first-class scientific discovery, the belts of charged particles trapped in the outer atmosphere by the earth's magnetic field, now named for the astrophysicist James Alfred Van Allen.

Replacement of tubes by transistors could not be pressed with the same speed on all fronts. In low-power, low-frequency devices such as broadcast receivers the transistor soon became ubiquitous (and transistor receivers in public places often obstreperous). Other fields were less easily conquered, but before long tubes were being pushed out of all applications except those involving high power or the highest frequencies, such as radio transmitters and radar. Electronic computers were among the principal beneficiaries. The size, cost, heat generation, and unreliability of even a moderately complex computer employing tubes were so great that by 1960 computers could have well become a case of arrested development, like cubism or Marxist theory, if it had not been for the transistor. The miniaturization afforded by the transistor was accompanied by two other developments that carried the process even further, to "subminiaturization": a corresponding decrease in the size of the associated circuit components; and the replacement of slow and unwieldy mechanical switches by tiny, fast-acting solid-state elements.

The first of these developments had led to what are known as *integrated circuits.* At about the same time that transistors began to make inroads on vacuum tubes, a technique for replacing circuit components such as resistors, capacitors, and the connecting wires by strips of various materials "printed" (actually, deposited by various processes such as electroplating, evaporation, and chemical etching) onto an insulator base was adapted to electronics. A stage—say, an amplifier or an oscillator—consisting of a transistor surrounded by a printed circuit took up a small fraction of the space needed by a vacuum-tube amplifier or oscillator with conventional components. The printed circuit is first drawn, much enlarged, on a sheet of drafting paper; the drawing is then photographically reduced and projected onto a sensitized surface; and after several additional steps a wafer of postcard size (and not much thicker) results to which transistors and other components may be affixed. How far could the photographic reduction be carried? Transistor manufacture had left the point-contact transistor far behind as new techniques, suitable for mass production, were developed. Before long, it became possible to reduce the active portion of transistors to a thin chip of semiconductor material only a few square millimeters in area. Now there was some point to reducing the associated circuits even further, below postage-stamp size, until the chip *and* its circuit were smaller than an earlier version of the transistor alone. Integrated circuits, some made up of several transistor chips and all the interconnections and associated components, are now made as units, usually by almost entirely automatic processes, and are bought by the makers of equipment such as computers, who are thus saved the trouble of assembly and testing. Whole electronic stages have become mere components.

The second important development, leading to what some enthusiasts dubbed "sub-subminiaturization," was in switching. Even before computers, automatic systems such as dial telephone exchanges required uncounted special switches known as *relays*. A relay is a switch operated by an electromagnet, usually a core of magnetic materials surrounded by a coil of wire. Current flowing through the coil raises a metal bar tied to a spring; when no current flows, the bar drops and completes a circuit. The main circuit is thus switched "off" or "on" according to whether a current flows or not in the auxiliary coil circuit. Early electronic computers of even moderate complexity contained hundreds of such electro-mechanical relays. The time needed to switch a relay on or off is about one-hundredth of a second. In the modern computer, millions of relay operations are performed by tiny *cores*, each made of magnetic powder that has been pressed into a millimetric ring and baked in an oven. Each core can be magnetized by a current along a wire strung through the center. The magnetization is oriented clockwise or counterclockwise according to the direction of the current, and the orientation can be detected by a second wire. Here again, the invention of a simple component that takes up very little space and cannot wear out came just in time to bypass an insurmountable obstacle in computer development. The mass production of both of these triumphs of solid-state electronics, transistors (or rather integrated circuits containing transistors) and magnetic cores, depends heavily on advances in materials science, notably techniques for controlling the purity of the basic materials. In an even more recent development, cores are being replaced by tiny transistor-like switches in integrated-circuit form, thousands to a single "chip," reducing space and power requirements and simplifying manufacture even further.

ELECTRONIC COMPUTATION *is the technique of processing information by electronic means.* There are two basic types of computers: *analog*, such as the slide rule and its electronic equivalents; and *digital*, such as the abacus, the mechanical desk calculator ("adding machine"), and the electronic digital computer. Analog computers work by measurement; digital computers, by counting. It is the ability of the electronic digital computer to perform complicated calculations at great speed that has made it the outstanding machine of the Second Industrial Revolution, an artifact that far eclipses any other in significance. The brief account of the history and operation of the computer that follows is therefore largely restricted to the digital computer.[4]

The problem of carrying out tedious calculations, such as tables of logarithms and of trigonometric functions, had fascinated mathematicians for centuries. At least two of them attempted to build calculating machines in the 17th century: Blaise Pascal (1623–1662) and Gottfried Wilhelm von Leibniz (1646–1716). Taking into account the then prevailing state of technology, it is not surprising to find that both men, who are also of towering stature in the history of philosophy, were great optimists; Leibniz, who also (independently of Isaac Newton) invented calculus, was later satirized for his belief in "the best of all possible worlds" as Dr. Pangloss in Voltaire's *Candide*. But Pascal did invent a method of mechanically "carrying" during addition, the ratchet gear by which a wheel containing ten numerals advances the adjacent wheel by one numeral after each complete revolution; and Leibniz conceived a machine that could not only add but also multiply. Neither machine was very practical. Not even the English mathematician Charles Babbage (1792–1871), who attempted to build a more sophisticated machine (he called it the Analytical Engine) nearly two centuries later, could overcome the mechanical limitations of his day. Apart from practical difficulties, Babbage's ideas were entirely sound. Among other features, he meant to use punched cards as input. Such cards had been used earlier by Joseph Marie Jacquard (1752–1834), the inventor of a loom (chapter 1) that was "programmed," as we would say today, by the cards to weave patterns. (Music boxes and other mechanical instruments, from player pianos to merry-go-round organs, work on similar principles.) Punched cards were first successfully used to process numerical data in 1890 by Herman Hollerith (1860–1929), whose calculating machine handled that year's census of the USA in a quarter the time that had been needed for the 1880 census.[5] At about the same time, William Seward Burroughs (1857–

1898) brought out the first recording adding machine to be successfully marketed and the National Cash Register Co. introduced the cash register, both crank operated. Millions of the descendants of these machines, long since electrified and otherwise improved, remain in use and have only recently begun to be supplemented by electronic desk computers.

The electronic digital computer was almost entirely an American development. Its first 20th-century ancestors were the mechanical computers designed by a group at the Massachusetts Institute of Technology during the two decades beginning in about 1925. The group was led by Vannevar Bush, the electrical engineer who later directed the U.S. Office of Scientific Research and Development during the war of 1939–1945. The MIT machines were analog computers (Bush called them Differential Analyzers) in which mechanical parts such as gears, levers, and cams were actuated by electrical parts such as motors and relays. Unlike an ordinary office-type calculating machine, in which the user normally intervenes after each operation, the Differential Analyzer was the first machine that could be "programmed" to carry out a series of operations without intervention, for instance to take a column of figures, multiply each by some number, and print out the column of answers. An improved model completed in 1942, in which the angle through which a gear had turned was measured electrically, saw extensive wartime use for the computation of artillery firing tables, conceptually a relatively simple problem involving millions of tedious calculations.

In 1937 Howard Aiken, a graduate student in physics at Harvard University, ran into a problem that countless scientists had met before him: the numerical solution of a mathematical expression called a partial differential equation, involving the interaction of several variables changing simultaneously. (A simple example is the thrust a rocket needs to develop at each point of its trajectory as it progressively uses up its fuel and grows lighter; a complex example is the effect of an increase in steel prices on the rest of a nation's economy.) Such expressions could be worked out only by making approximations, and even then the computation was often so lengthy that many a solution had to be given up as intractable. But Aiken decided to do something about it. With the help of four engineers at the International Business Machines Corp., which underwrote the project, he set about to devise and construct the first fully automatic digital computer. The IBM Automatic Sequence Calculator, Mark I, began operating at Harvard in 1942, a short walk from MIT's Differential Analyzer, which it resembled in consisting of mechanical parts operated electrically, but whose capacities it greatly exceeded. IBM had taken up Hollerith's idea of using punched cards, and the Mark I was controlled by their equivalent, punched tape, by which instructions laboriously prepared beforehand were fed into the machine. It could complete multiplications involving up to 23 digits in seconds. The Mark I was the first realization of Babbage's dream of an Analytical Engine; it had taken technology the best part of a century to get to a point where the mechanical arts (with a bit of help from the electrical) were equal to the task.

The stage was now set for an electronic digital computer, in which the relays and other electro-mechanical devices would be replaced by the much faster-acting electron tubes. Such a machine, the Electronic Numerical Integrator and Calculator (ENIAC), had in fact been under design for some time at the University of Pennsylvania, where a hundred operators and a Bush Differential Analyzer were kept busy computing tables for the U.S. Army. (The U.S. Navy had in the meantime ordered a Mark II from Aiken; and George Robert Stibitz and others at Bell Laboratories were also experimenting with relay calculators.) The army had encouraged the Pennsylvania team's proposals, made jointly by the physicist John William Mauchly (1907–1971) and the electrical engineer John Presper Eckert, Jr., for an electronic digital computer, in the hope of replacing the Bush Differential Analyzer by a much faster calculator; but the war had ended before ENIAC was delivered. It did work much faster, since a vacuum tube when used as a relay has a switching speed of less than one-millionth of a second (a microsecond). ENIAC was also tremendously complex. It contained 18,000 vacuum tubes, at least one of which failed, on the average, every few hours. Instructions had to be given by hooking up special external circuit boards, not unlike portable telephone switchboards, some of which took days to prepare. Nevertheless the machine lasted for nearly 10 years and ran up thousands of hours of operating time.

The evolution of the high-speed digital computer from the primaeval ENIAC was a scientific and technological achievement of the first order, the more so since it was accomplished in about 15 years. To be sure, ENIAC used existing art—there was little in it that could not have been done just as well in 1936 as in 1946 if the war had come sooner. Its speed (roughly, the number of multiplications per second) was about 300, a thousandfold increase over that of its electro-mechanical predecessors. Its "memory" or storage—the capability of recording intermediate data in a computation of several steps—consisted of 20 numbers.

It was perfectly obvious to the engineers that they could do much better if the storage could be enlarged and, equally important, if the logical structure of the machine's operation could be improved so that hooking up the external circuit boards before a series of calculations would be unnecessary. Enlarging the storage without increasing the "access" time (the time required for the central processor that performs the computations to communicate with the memory) was an engineering problem that was ultimately solved by the solid-state electronic components described above, transistors and magnetic cores. Before they came along, some fairly esoteric solutions were attempted, deriving in part from wartime radar technology.

The other development, improving the logical structure, was solved in the main by the great mathematician John von Neumann (1903–1957), using in part the ideas of another mathematician, Alan Turing (1912–1954). Von Neumann conceived the "stored program," a rearrangement by which instructions as well as data are stored in the machine internally. This feature was included even before transistors were introduced, for the first time in an English machine put into operation at Cambridge University in 1949, and then in a number of American experimental computers. The first commercial machine with a stored program was the Remington Rand (later Sperry-Rand) Company's Universal Automatic Computer (UNIVAC I), designed by Eckert and Mauchly and delivered to the U.S. Census Bureau in 1951. It is now in the Smithsonian Institution in Washington, D.C.

These machines were capable of another tenfold increase of speed over ENIAC, to 3000 multiplications per second, and memories that were accessible in microseconds. Further computer development accompanying the introduction of the solid-state active and storage components in the 1950s led to multiplying speeds approaching a million per second. That development is still continuing; the end is not in sight. Simultaneously with giant machines of increasing capabilities came the astounding proliferation of smaller, less expensive special-purpose computers, ranging from aids for scientific laboratories down to quite simple desk computers, smaller but more versatile than the mechanical calculators they are replacing.

An even more important part of the further development has been the continuing elaboration not only of the machines themselves but of the method of interconnecting and programming them, so much so that many computer experts need only a sketchy notion of what is inside the machines and spend most of their time on devising new ways of using them, sometimes inventing special computer "languages" for special applications. As to the physical advances, most spectacular has been the improvement in presenting the information, the "printout," which no longer necessarily takes the form of a sheet of digits, but avails itself of a whole arsenal of graphic devices, from cathode-ray tubes to permanent displays ("hard copy") of very high quality; a perfectly drawn figure, ready for insertion in a printed text, is quite readily obtained and alphanumeric displays, for instance for book printing, can be reproduced at astonishing speeds—tens of thousands of lines per minute.

One feature of the computers that came after ENIAC was employment of the binary system, a method in which only two digits are used for arithmetic operations in place of the ten digits of the decimal system. All numbers are represented by combinations of 0 and 1, so that the decimal-system numerals from 0 to 9 become 0, 1, 10, 11, 100, 101, 110, 111, 1000, and 1001 in binary. That a three-digit decimal-system number such as 365 has nine digits in binary, 101101101, does not slow down computations; on the contrary they are made much easier by the fact that simple binary circuit components can be employed. A switch or a light can be off or on, a pulse of transistor current can be present or absent, a magnetic core can be magnetized clockwise or counterclockwise, a specific spot on a punched card can have a hole punched in it or no hole. These are all examples of binary or two-state elements, faster, cheaper, more reliable, and more easily put together in a logical arrangement than components that can show more states.

Another feature of modern computers is what might be called their decision-making ability. Faced with two possible actions, a computer can be programmed to take one or the other depending on the outcome of a previous action, for instance on whether the result of a calculation just made was above or below a certain value. As a simple example, in a payroll calculation the computer is programmed to multiply each employee's hourly wage by the number of hours he worked in the past week. Say the employee's contract calls for 35 hours of work per week at a certain wage, with half again as much for overtime. Then if the hours he has worked add up to 35 or less, the programmed multiplication yields the gross wage. If they add up to more than 35 hours, the excess hours must be further multiplied by 1.5, to yield the pay at the higher overtime rate, and the result added to the regular pay. The computer must be programmed to decide whether to make this further calculation or to go on to the next employee, according to whether the number of hours worked adds up to more than 35 hours or not. (By a similar process, it can also decide whether or not income tax should be deducted, subtract any applicable insurance payments and other deductions, and print the pay check.)

Finally, a computer can adapt its store of knowledge automatically as a result of information that it has itself generated. In 1962, Arthur Lee Samuel, an electrical-engineering professor who had gone to work for IBM, programmed a computer to play checkers games of championship quality by storing in its memory sequences of moves that had successfully captured the opponent's pieces and reached his king row in previous games. The machine became a better checkers player as it went on. It could estimate the effects of various possible plays for several moves ahead—the ability that distinguishes the champion player from the amateur—and decide, with increasing probability of success, which move to make next. This adaptive ability, the ability to change the stored information as the computation goes on, makes it seem that the computer can learn or understand, in other words, exhibit intelligence; and in fact the term *artificial intelligence* has been coined to describe this property.

Opinion is divided on whether computers can be said to "think." If intelligence is defined as depending on consciousness, computers cannot have it: no computer is conscious of itself. But defining intelligence on the basis of human attributes is fatuous; it is a little like defining "thinking" as whatever computers cannot do—a comfortably elastic definition that remains endlessly adjustable in the face of further progress. One should perhaps view such attitudes in the context of man's primordial fear of the existence of an intelligence greater than his own, a fear that is part of religions ancient and modern. Humanoid automatons that run out of control abound in legend and literature. We read of the robots (a Czech word meaning workers under compulsory labor) of Karel Čapek's 1921 play, *R.U.R.* (Rossum's Universal Robots), who take over the world; of the Sorcerer's Apprentice of Goethe's 1798 ballad (and Paul Dukas's symphonic poem), who does not know how to stop a water-carrying servant he has created from a broom; of the fantastic tales of E. T. A. Hoffmann (used by Jacques Offenbach and Léo Délibes for opera and ballet) about a singing and dancing doll and other automata; of Frankenstein's monster, created in 1818 by Mary Godwin, the wife of the poet Shelley; and its legendary ancestor, the uncontrollable clay Golem of Rabbi Löw in 16th-century Prague, an expression of the fantasies of an oppressed minority.[6]

The advent of the computer has done nothing to lessen mankind's fascination with the subject, as attested by the popularity of the plot in which the computer "takes over" in contemporary science fiction.[7]

Objective definitions of what is meant by intelligent behavior remain in short supply. Thus, a computer can evidently learn, it can respond to new situations, it can solve problems, it can direct other machines and even people, it can answer questions. But it can only do so in ways foreseen by the human programmer—though it is true that it can do it vastly faster, performing complex computations in a few hours that it would take a human several lifetimes to compute. Even the designers of computers capable of playing checkers and other games cannot match a memory that can keep all possible future moves in mind. But the "mind," the intelligence, is artificial, apparent, not real. Or is it? There is a test, proposed by Turing and known as the Turing Test. In such a test, an examiner would send questions to another room, from which answers could be sent back by a man or a machine; and the examiner would have to guess which by the answer. Only when the examiner could not tell the source of the answer could the machine be said to be thinking. There is no computer that could win Turing's Test—and it would seem there never will be. Turing himself proved a theorem showing that there must be critical questions that computers cannot be programmed to answer; when asked such an "undecidable" question, the computer would give either a wrong answer or no answer. But that did not lead him to downgrade artificial intelligence in comparison with human intelligence. Before his tragic death in 1954 at the age of forty-two, he said the following in what has become the classic article on "Can a machine think?"

> Whenever one of these machines is asked the appropriate critical question, and gives a definite answer, we know that this answer must be wrong, and this gives us a certain feeling of superiority. Is this feeling illusory? It is no doubt quite genuine, but I do not think too much importance should be attached to it. We too often give wrong answers to questions ourselves to be justified in being very pleased at such evidence of fallibility on the part of the machines. Further, our superiority can only be felt on such an occasion in relation to the one machine over which we have scored our petty triumph. There would be no question of triumphing simultaneously over *all* machines. In short, then, there might be men cleverer than any given machine, but then again there might be other machines cleverer again, and so on.[8]

The debate continues. It ranges wide: in Norbert Wiener's *God and Golem, Inc.* (1964), the author considers the religious and ethical implications of the confrontation between man and machine, and cooly advises us to "render unto man the things which are man's and unto the computer the things which are the computer's." The argument is no longer restricted to whether machines can think like humans. Can they also manipulate symbols at random to create works of art—music, paintings, poetry? Can they develop original knowledge beyond the human brain's capacity to conceive? May they not come to constitute an "artificial intelligentsia" capable of dealing with higher abstractions than man can and thus forge into the unknown? It is in that direction, the development of the computer as a tool in furthering man's own evolution, that the most exciting challenge lies; not the possibility that the computer will go on an uncontrollable rampage. That is not to say that we should consider all possible computer applications as harmless. On the contrary great vigilance is required to avert misuses. But neither is it necessary to swallow whole the alarmist cries of some journalists (who seem to take their cue from overenthusiastic computer salesmen and outright charlatans on the fringes of the artificial intelligentsia) that computers will "take over." They will not. As computer designers are given to saying, "You can always pull out the plug!"

ELECTRONIC CONTROL INVOLVING COMPUTERS *is the logical combination of the methods of industrial electronics (chapter 3) and computer technology.* The principles of feedback and servomechanisms mentioned earlier are of primary importance. The addition of the electronic computer makes for great speed and sophistication in data processing. We have already mentioned a task, keeping a space vehicle to a prescribed trajectory, which could not be carried out without computer control, since the myriad decisions and corrections that must continually be made would not be possible without it. Even in an ordinary aircraft such decisions can no longer be left to the unaided judgment of the pilot. At the perfectly safe height of a few hundred meters, a jet aircraft flying well below the speed of sound need not dip its nose onto a trajectory too far below the horizontal—a few degrees is all it takes—before it finds itself on an irreversible collision course with Mother Earth; the time needed for the pull-up maneuver, even assuming the pilot could start it without delay, is greater than the time left before the crash.

The first attempts to introduce what computer engineers call "on-line, real-time" control (the nearly instantaneous control of a process while it is under way) did come in aviation, not in an attempt to keep aircraft aloft but to shoot them down. In this problem, which arose early in the war of 1939–1945, the control mechanism imitates the hunter of a moving quarry, who must aim ahead of ("lead") the target to allow for its motion during the time the arrow or other projectile will need to reach it. But if the target is under the control of a man—the enemy pilot—capable of taking evasive action, it can change its expected position before the next projectile gets there, so that prediction seems impossible. In fact such prediction *is* possible, for two reasons. First, the choice of maneuvers open to the pilot is limited by aerodynamic, physiological, and psychological considerations. Second, the problem can be treated statistically, with good probability of success.

From a series of more or less exact readings taken on the aircraft's position, for instance by radar, one can guess where the aircraft will be at some future time, even if some of the readings are slightly in error and even though the flight path is not a straight line. In making a prediction from such measurements distributed in time (corresponding to what statisticians call a "time series"), one gives less weight to older data than to the more recent; and not simply less weight in direct proportion to the age of the data ("linear" prediction), but following a more complex, "nonlinear" weighting scheme. For instance, one can keep checking on how successful the older data are compared with the new in making accurate predictions and go on adjusting the weighting scheme accordingly from instant to instant; or one can evolve some other scheme of nonlinear programming that goes on improving the estimate of the future.

A linear prediction theory and the theory of several specific nonlinear predictors were worked out by Wiener in the early 1940s. It was soon realized that the theory (which had been independently developed at almost the same time in Russia by Andrei Nikolaevich Kolmogorov) was mathematically related to the solution of a more general problem in communications, the detection of a weak signal in the presence of interference. We find it difficult to conduct a conversation over a noisy telephone line (or even at a large cocktail party) or to follow a shortwave broadcast in the presence of fading and "static," especially if the message is coming across in a foreign language that we understand imperfectly. (Another example is a faint radar return from a distant target.) The problem of detection of the true signal among the noise turned out to be closely related to the problem of the prediction of its value at some future time. In both problems, the concepts of smoothing and filtering are of great importance, concepts that had occupied telephone and other electrical engineers for some considerable time. In fact the publication of Wiener's great work *Cybernetics* in 1948 was preceded by a classic paper on "Theory of Communications" by Dennis Gabor, an electrical engineer, in 1946, and closely followed by *A Mathematical Theory of Communications* (1949) by Claude Elwood Shannon of the Bell Telephone Laboratories.[9] This new branch of learning, *communication theory* (or information theory as it is often called), firmly based on the twin foundations of mathematics and electrical engineering, is not so much concerned with detecting or predicting signals distorted by noise as with a complementary problem: given a noisy transmission channel, what is the best way of "encoding" the signal so that it comes across with least likelihood of being distorted?

That problem is as old as electrical communications. It occupied Morse as early as 1838, when he devised the code that bears his name. The problem of assigning combinations of dots and dashes to letters of the alphabet was solved in a simple and ingeniously straightforward way. The letters that occur most frequently in English, *e* and *t*, got the simplest symbols—a dot and a dash—followed by *a* (·–), *n* (–·), and so on up to combinations of four dots or dashes for the rarer letters. How often a letter would come up was estimated from a count of the types available for each letter in a printer's box of type. This simple procedure was surprisingly effective. Modern communication theory shows that relatively little would be gained—about 15 percent—by the use of the ideal, theoretically most efficient assignment possible. But until the development of the theory, there was no way even to state the problem, let alone to optimize it. The pioneers of electric telegraphy had intuitively grasped the main postulate of communication theory; it is extremely important how one goes about translating information into electrical signals. Devising the Morse code is one of the simplest problems of communication theory. When it comes to more complex tasks of transmitting patterns such as speech, music, pictures, computer data, or satellite photographs of the surface of Mars, the way in which such information is encoded and the nature of the transmission channel are clearly of the most profound importance.

The origins of cybernetics are thus seen to be rooted in such problems of communications as encoding and decoding, smoothing and filtering, detection and prediction, and the nature of the transmission channels. It was Wiener who pointed out that problems of feedback control and servomechanisms likewise belonged to communication engineering, not to power engineering as had been assumed, and who helped launch a unified treatment of these fields. Electronic computers were also fitted into the communication framework, regardless of whether they were being used for switching telephone calls or controlling automatic fabrication.

It is in the employment of computers in business and industrial processes that cybernetics has had the greatest impact, going beyond merely technical considerations. The automatic factory in which human intervention is reduced to the setting up of the successive steps in manufacture and to maintenance work, the bank that immediately posts all the transactions of its every branch and enters them in an up-to-date central record, the wholesale firm which adds to its inventory on the basis of a continuous monitoring of what it has just sold, the airline which can instantaneously assign a seat on any flight at the request of a clerk at any one of thousands of ticket counters—all these are instances of industrial processes in which the computer is the all-important component. Its countless other actual and potential uses—in science, in education, in weather forecasting, in medicine and other arts—make it abundantly clear why the computer deserves to be singled out as the most important artifact of the Second Industrial Revolution.

A corollary is that this tremendously important development is not without dangers, deriving from what most would regard as misuses of computers. Invasion of privacy by large data banks, the potentialities of taking away control from the many to give it to the few, the massive deployment of computers in war, and the unnecessary introduction of computers into problems better solved in other ways are examples of problems that industrial societies must face while taking advantage of this marvellous new tool.

When 100 laboratory workers were asked which of two unlabeled designs they preferred and which they thought was by Piet Mondrian, 59 preferred engineer A. Michel Noll's quasi-random pattern generated by means of a computer (*right*); only 28 correctly identified the black-and-white reproduction of "Composition with lines," 1917 (*left*), as Mondrian's painting. (*IEEE Spectrum*, 1967)

Chapter 5 Some Aspects of Contemporary Technology

ENERGY *and new methods of harnessing it are among the chief earmarks of contemporary technology.* Primitive man depended on his muscle power alone until he learned to use domestic animals, running water, the wind, and fire. The Industrial Revolution added first steam and then electricity, which had been known for some time but had not been used to provide motive power, and also chemical combustion. World energy consumption rose at an ever-increasing rate, to the point of arousing concern over known reserves of the main source of energy, fossil fuels—oil, natural gas, and coal, which between them accounted for 95 percent of the energy supply until recently—in the foreseeable future. The prospect might brighten somewhat as new materials make operation of power plants at higher temperature possible, which in turn would make higher efficiencies possible; at present, about two-thirds of the total energy used is lost in its thermal conversion. Efficiency can unfortunately never be 100 percent—according to the Second Law of Thermodynamics, no heat engine can fully turn all the heat into work.

Among other phenomena known to science that await technological application are thermoelectricity, photoelectricity, chemical conversion, and nuclear fission and fusion. Various stages of development have been reached and many difficulties remain. Because of their urgency, the problems of putting these processes to efficient use occupy some of the best engineering brains throughout the world.

Thermoelectricity means the direct conversion of *heat* into electricity, without the wasteful intermediate step of first producing motion (as in a steam-operated electric generator). A simple example is the thermocouple battery. If two strips of different metals are joined at one end and the junction is heated, a voltage appears between the other ends. The device is widely used as a sensitive thermometer; its efficiency as a power converter is still extremely low—below 10 percent.

Photoelectricity, the direct conversion of *light* into electricity, is familiar to anyone who has ever used a light meter in photography. Owing to the demands of the space age—to provide continuing power for communication satellites and other space vehicles—this method has seen considerable development. "Solar batteries" for converting sunlight into electricity have reached the relatively high efficiency of about 15 percent. Somewhat higher efficiencies will open up many new applications, ranging from powering individual households to distilling potable water from the ocean. Generation of electricity is only one of the uses of solar power. It can be also used for direct heating of houses and it can be recovered by utilization of the natural photosynthesis that occurs in plant growth. (The plants can then be turned into fuel, for instance by fermentation into alcohol.) Neither storage of solar heat nor photosynthesis is very effective at present, but energy from the sun is free and plentiful in many parts of the world, so that even an inefficient method may be economically significant.

Another kind of battery is the fuel cell, in which energy from a *chemical* reaction appears in the form of electricity rather than heat. The fuel cell resembles the storage battery used in cars, except that the fuels (which may be gases such as oxygen and hydrogen) flow steadily through the cell. A particularly attractive application for the fuel cell would be as replacement of the internal-combustion engine in cars (chapter 2), since electric cars produce little air pollution or noise. But the gasoline engine has become so effective during the many decades of its development that it is difficult for any alternative to compete with it. Moreover, cleaner and quieter gasoline engines and turbines will doubtless be forthcoming as a result of legislation based on public demand.

The most promising new source of energy is *nuclear* or "atomic" energy, which results from one of two processes: fission or fusion.

In *fission*, the nuclei of the atoms of the nuclear fuel (usually uranium or plutonium) are penetrated by high-speed particles that break up the nuclei (in popular parlance, the atoms are "split"), which makes additional particles available for splitting other atoms (we speak of a "chain reaction"). The fragments constitute elements of lower atomic weights—the scientists have achieved the age-old dream of the alchemists to transmute one element into another—but the total mass of the fragments is less than the original mass and the rest has become available as energy in accordance with the famous law of Albert Einstein (1879–1955) relating energy and mass, $E = mc^2$. Since c is the speed of light, 300,000 km/sec, c^2 is a very large number and the resulting energy is also great: the conversion of a small fraction of 1 kg of uranium releases as much energy as is obtained from burning nearly 3000 metric tons of coal. The energy (first released in the "atomic" bombs of 1945) appears in the form of heat, which can be used to produce steam for a turbine and thus electric power—much as in a fossil-fuel steam plant, except that much less fuel is required: a nuclear reactor can run for years on a single fuel charge. Many such reactors are currently in operation. In the USA, they have already come to dominate all new electric-plant construction and have also been successfully employed in naval vessels. But the operation of nuclear reactors poses many a problem: they require exotic materials and heavy radiation shielding (which makes them unsuited for cars or aircraft), they operate best at temperatures so high as to corrode most ordinary materials, and they produce radioactive wastes that require careful storage.

Because of these difficulties, the major atomic-energy laboratories of the world are all trying to utilize a process that is the opposite of fission, called *fusion* or thermonuclear reaction. In that process (likewise first artificially achieved in an explosion, the "H" or hydrogen bomb), two light nuclei fuse to produce a heavier one, but not as heavy as the sum of the parts, so that the excess mass again becomes available as energy. This process goes on all the time in the sun, but to bring it about on earth, the nuclear fuel (usually deuterium, an isotope of hydrogen) must be held at the sun's temperature—about 100 million degrees C—long enough for the thermonuclear reaction to take place. This condition has been met so far only in the hydrogen bomb, in which a fission-type atomic bomb is used to "trigger" the fusion reaction and an uncontrolled release of energy—the H-bomb explosion—takes place. To harness this energy, that is, to produce a controlled thermonuclear reaction, is the dearest wish of hundreds of scientists and engineers the world over. The technological payoff would be immense, since deuterium is plentifully available and the process leaves only modest amounts of radioactive wastes. (On the other hand, the containment structure is very radioactive and some of the materials are hard to handle.)

Apart from the possibility that an entirely new and hitherto unknown energy source may turn up, it is likely that in the short run some combination of the above methods may prove to be the most effective. It may be that the most efficient utilization of nuclear reactors will come when they are "topped off" by the thermoelectric devices that can take full advantage of the high reactor temperatures; and other combinations are also feasible. Meanwhile, back at the atomic laboratories, efforts will continue to insure that the world does not end with a bang but continues to run with a whimper—the whimper of controlled thermonuclear reactions.

THE NEW FOOD REVOLUTION, *technology's answer to the problem of growing more food of the kind that is needed in the places it is needed most,* is a fairly recent phenomenon.[1] Before overpopulation can be throttled, highest priority must be given to increasing food production, especially in developing countries. The specter of a world food crisis has led to a new emphasis on agricultural investments in raising and processing farm products.

Most developing countries lack facilities for agricultural research and higher agricultural education of the sort long taken for granted in most industrialized countries. But there are some notable exceptions. Mexico, for years an importer of half the wheat consumed in the country, became self-sufficient in that grain within 12 years after an extensive wheat research program begun in 1944 with Rockefeller Foundation help, and by 1964 exported half a million tons. In the Philippines, research jointly sponsored by the Rockefeller and Ford Foundations developed two new varieties of rice, the IR 5 and IR 8, which yield fully 15 times the average yield of traditional varieties; within three years, the Philippines were transformed from a rice-deficit to a rice-surplus country. In Indonesia and Thailand, agronomists from the USA produced something of an agricultural revolution by adapting Guatemalan corn to local conditions; in Thailand, the corn crop rose in eight years from virtually zero to a level that produced a $104 million export, the fourth largest in the world.

Extended to other critical food-deficit areas, the results of these largely privately financed research efforts have produced what has come to be known as the "green revolution" in some of the world's food-deficient regions that not long ago had been threatened by widespread and seemingly continuing famine. In India, the area planted with new varieties of wheat, rice, millet, and sorghum increased from about 23,000 acres in 1965–1966 to nearly 4 million acres just one year later; five years later, the Indian harvest was over 100 million tons and self-sufficiency in these foods had been achieved. By introducing their Taiwan Native 1 rice species in Africa, development teams from the Republic of China (Formosa) more than doubled the crop from traditional rice strains in the Cameroons, Gabon, Gambia, Malawi, Niger, Senegal, Chad, and Togo. As a result of these and other developments, farming has suddenly become both respectable and lucrative in many countries.

The agricultural revolution is not confined to progress in developing countries. New and as yet untried ideas will displace a great many outdated notions about food production—first in the advanced countries, but eventually with profound effects on the developing countries. For the first time in recorded history it may become possible to feed, and feed well, all mankind, even taking population increases into account. Predictions for the next generation, based on a recent study of agriculture in the USA, include:

Wheat harvests of 300 bushels per acre, more than 10 times present yields.

Corn outputs of 500 bushels per acre, compared with the present average of 75 bushels.

As many as 1000 calves during a single cow's lifetime (as against the present 10), through the use of hormone control of the ovulation and pregnancy cycles of cows and other domestic animals; embryos will be transferred from special breeder animals to ordinary "incubator" animals, with substantial increases in a herd's prolificacy and quality.

Milk production more than tripled and supplemented by synthetic milk made from carrot tops and pea pods.

Replacement of farm laborers by agricultural engineers trained in fields ranging from electronics to air conditioning and capable of designing automated and computerized farms to a degree at which tilling of the soil, planting and harvesting crops, and regulating the growing process will be controlled from a center equipped with computers, radar, and remote-control devices.

Fields covered by huge plastic domes or other means of maintaining constant control of the environment.

Remote-control harvesters that not only pick but also grade, package, freeze, and deliver food to wholesale depots.

Space satellites collecting up-to-the-minute, worldwide data on soil conditions, insect damage, and crop growth patterns, and making long-range weather and production forecasts.

Investment in agriculture in developing countries is bedevilled by a nontechnical problem: the shifts resulting from the concentration of production, capital, and manpower that follow large-scale investment. Another, more political problem is that foreign owners of plantations may be accused of colonialism, a politically explosive charge. But there are other attractive investment possibilities: in supplying the new varieties of seed, fertilizers, insecticides, water and farm machinery, and economic services such as credit and marketing facilities—not to speak of large-scale irrigation, rural electrification, storage, and transport. The Food and Agriculture Organization of the United Nations has made a special effort to mobilize the resources of international business firms for food production in developing countries through the FAO/Industry Cooperative Program, a joint effort to establish agriculture-related industries in developing countries.

In addition to producing fertilizers and farm machinery, international business is paying increasing attention to novel methods of producing proteins from petroleum and natural gas, and from various vegetable and marine sources. Apart from synthetic nutrients, the oceans may well become a major source of food in the future, not only by an increase in the fish and shellfish catch both for direct human consumption and for indirect consumption in the form of fish meal and other products processed for animal feeds, but also by "ocean farming"—the raising and harvesting of sea animals and plants in shoreline ocean beds and on ropes suspended from special rafts. The word "aquaculture" has been coined for such endeavors.

In the developed countries, food production is already one of the world's great success stories. This great achievement of the last two decades can be attributed to reliance, not on grandiose theories, but on treating food production as a business from the farm to the processing stage and distribution, all requiring skills and capital inputs essential to ultimate success. If the psychological and other nontechnical problems can be overcome as well as have been the technological, there is every reason to hope that the world will be well fed in the future. Through the close cooperation of governments and private business of all nations, the coming decades can go down in history as the age of the revolution that achieved this monumental feat.

MATERIALS MADE TO ORDER, *notably the artificial combinations of metals (alloys) and resins (plastics) specially developed for particular applications*, are another outstanding characteristic of modern technology. In the past, materials were used essentially as they were found and extracted, with relatively few changes in their chemical composition. With the advances made in materials processing during the 19th century (chapter 2), it became possible for the first time to create materials to specifications, not only for such esoteric uses as drugs and dyes, but also in bulk applications in construction, transport, clothing, packaging, and dozens of other industries. The trend toward fashioning materials to specific applications is nowhere so well exemplified as in the field of plastics.

During most of the first century after Hyatt invented celluloid, plastics served as substitutes for ivory, guttapercha, rubber, natural fibers, and a host of other natural materials. The somewhat contemptuous German word for "substitute" (food or material), *Ersatz*, achieved general currency during the period of worldwide shortages brought on by the war of 1939–1945. A mildly pejorative meaning still attaches to the word *plastic* as a cheap and therefore inherently inferior substitute for a "real" (i.e., natural) material, an image that today's plastics industry is striving hard to replace by another: as the material that makes entirely new developments possible—the stuff that dreams are made on.

To some extent, this view is not unjustified. Although the low-cost, "throw-away" uses are the most familiar to consumers even in poor countries, the designation "plastic" has gradually achieved something closer to its actual technical meaning: a man-made, deformable polymer (i.e., compound whose chemical structure is a long chain of repeating groups of molecules), with properties often tailored to specific applications, many of them first made accessible to mass production precisely as a result of the availability of an inexpensive raw material. Conversely, when an industrial designer replaces a natural material such as wood or metal by a plastic, he usually does more than simply substitute one for the other. The real payoff often comes when the product is redesigned to take advantage of the specific properties of the new material in improving such characteristics as machinability, performance, and reliability while reducing cost. Part of the cost reduction is sometimes the result of simplified finishing, since few plastics require elaborate polishing and similar operations. Thus, a product made from a plastic material that costs $1 a kilogram may turn out to be cheaper than the same article made from a metal priced at a quarter the bulk cost.

Strictly speaking, any substance is plastic if it is deformable. All plastics are deformable in one sense or another—if not in their solid finished form, at least during the forming process. Some plastics are supplied in liquid form and do not become solid until used, as in an adhesive or a paint. Solid plastics may be rigid (for instance a building material), flexible (a fabric), or elastic (a car tire).

They may be also classified according to use and cost. Commodity plastics are the cheapest; they are used to produce most of the familiar, visible consumer products. Engineering plastics, priced at a few dollars per kilogram, go into technical products such as industrial applications. Exotic plastics, which can cost $50 per kilogram, have special properties such as good resistance to atomic radiation or to great extremes of heat and cold, needed in nuclear physics or space applications.

Still another classification distinguishes among ordinary solid plastics and materials that form a film, are capable of being blown up into a low-density (but relatively strong and stable) foam, or are combined into composites or reinforced plastics such as fiberglass (which is only about one-third glass).

But the most basic distinction relates to one physical property: behavior under the action of heat. Thermoplastics melt under heat and harden with cooling, a process that can be repeated over and over again. Thermoset materials, once melted and hardened, retain the shape given to them the first time. This property also governs the method by which plastic products are manufactured.

Thermoplastic products are made in the main by four processes: (1) *injection* molding, in which melted plastic powders or pellets are forced into a hollow mold to make a solid piece; (2) *thermoforming,* in which a plastic sheet is clamped over a mold and pulled in, usually by a vacuum, to form a hollow product such as a box; (3) *extrusion* of sheets, rods, pipes by force through a mold of the desired cross-sectional shape; and (4) *blow molding* (similar to glass blowing), another process by which hollow forms such as bottles are made, in which an air stream forces the material against the walls of two halves of a take-apart mold.

Most thermoset materials are processed by either compression, in which plastic pellets or pieces of dough are simultaneously subjected to heat and pressure in a mold; or transfer molding, in which the material is melted beforehand. Still other methods of fabrication are used to manufacture foam products and reinforced and composite themosets, depending on the nature and form of the reinforcing material.

The great proliferation of plastics fabrication techniques has led to the emergence of an entire auxiliary industry, the manufacture of machine tools and other devices by which plastic products are made, tested, assembled, and packaged. Molding presses, extruders, cutters, blowers, coaters, and many other machines are supplied to manufacturing plants ranging from one-man shops to factories employing thousands. Annual world production of such machines is reckoned by the tens of thousands.

Minor variants apart, at least 2000 plastic compounds are in industrial use. Production of the top five amounts to millions of tons a year. The most widely used in polyethylene, strong, flexible, resistant to chemicals, familiar to consumers in the form of packaging, bottles, toys, and dishes. Vinyls are next, used for raincoats, transparent containers, floor and wall coverings, and a thousand other things; one of that family alone, polyvinyl chloride (PVC), accounts for a substantial portion of total plastics production. Polystyrene is almost as ubiquitous, moderately strong, glassy in appearance, although also familiar in the feather-light foam form. Phenolics are fourth—and oldest—cheap, strong, and heat resistant, the material from which combs and car distributor caps and other electrical insulators have been made for over half a century. Number five is the formaldehydes: urea (buttons, appliance housing) and melamine (resistant table-tops, tableware).

The above examples of applications are drawn largely from consumer products, which are the most familiar. Actually, plastics have a great many uses, some of them much less obvious to the untrained eye. In recent years, the construction industry has become the largest customer for plastics. Structural and wearing surfaces, insulation and pipe, plumbing and lighting fixtures, and above all glues, coatings, sidings, and trims account for more "invisible" plastic elements than the typical new home owner realizes. In nonresidential buildings, plastics likewise compete with traditional materials in floorings, stacks, thermal insulation, window frames, and even the windows themselves.

Packaging is a growing industrial market for plastics, not only for food containers but also for shipping bags, crates, shrink wraps, and packing materials. Transport is another. Plastics are being used for cars and trucks, in gasoline tanks, headlight covers, radiator grilles, interiors, panels, and—in some models—car bodies. Small boats are major users of fiberglass-reinforced plastics: more hulls under 6 meters long are made of plastics than any other material. Even aircraft manufacturers are increasingly relying on plastics.

A field with great potential as a user of plastics is agriculture. Plastic films are used for seed packaging, greenhouse walls and roofs, and for protection of crops against insects, storms, frost, and loss of moisture (both as a cover and as a mulch). Drains, pipes, tunnels, and even small dams and reservoirs utilize plastic materials.

Another field in which plastics are catching on is medicine. Already items such as hospital utensils, surgical trays, and some medical instruments are made partly or wholly from plastics; so are casts and splints, artificial limbs, dentures, and replacements of biological materials generally, including skin grafts and cardiac bypass pumps. These and laboratory uses, present and prospective, are virtually unlimited.

One unresolved problem derives from the very success of plastics in resisting deterioration. A visit to a car wrecker's yard shows that among the rusting steel, tarnished chrome, and other materials the only parts that still look shiny and new are the plastic tail lights. The fact that plastics do not degrade easily is currently the subject of intensive research in the industry, which is under great pressure to deal effectively with the environmental problems of discarded plastics.

TECHNOLOGY AND THE HEALING ARTS *represent a coming together of quantitative and qualitative approaches toward a problem as old as civilization.* Use of technological tools, for instance as medical instruments, is nothing new, but the concerted effort to bring the entire range of technology to bear on the biomedical sciences is a more recent endeavor, dating back no further than the rise of the modern pharmaceutical industry in the mid-19th century. Advances in organic chemistry made it possible to isolate (and, in some important cases, to synthesize) the active ingredients of drugs and anesthetics that had been previously available only in vegetable form. These developments not only alleviate suffering by putting an end to the difficulties of prescribing correct dosages when the concentration and purity of most drugs were uncertain, but also made the resulting products available in vast quantities. Sulfa drugs, antibiotics, cortisone, and many other milestone achievements of pharmacology would have remained expensive laboratory curiosities if it had not been for the industrial development that made their mass production possible. Likewise, radiology, cardiography, encephalography, and a multitude of other technical contributions ranging from modern hospital construction to the design of dentists' chairs would have had very little impact if they had not been followed by the corresponding feats of electrical and mechanical engineering that insured their widespread adoption. Technology has also made innumerable other contributions to the practice of medicine, including more accurate diagnosis and more skillful treatment and the conquest of infectious diseases.

A parallel development has been the application of technology to the field of public health. Few opponents of technological advance would advocate a retreat to a simpler and presumably cleaner time in the past if they realized that it would mean a return to, say, outdoor privies. Even so prosaic an advance as the invention of the flush toilet at the end of the 18th century was a tremendous contribution to sanitary living; making potable water conveniently available was another. Before that, epidemics resulting from unsanitary conditions used to wipe out entire towns. Similarly, scourges such as yellow fever, malaria, typhus, smallpox, and poliomyelitis have virtually disappeared as the result of medical research accompanied by technological advances.

The Second Industrial Revolution uncovered so many ways in which engineers interested in the solution of biomedical problems could help their colleagues in the life sciences that a new profession was created—*bioengineering*. Its practitioners receive a combined technical and scientific education and work at a great variety of endeavors, ranging from purely scientific research (for instance, learning how cells behave from measurements of their electrical properties) to the more mundane tasks of designing entirely new kinds of medical instruments.

Instrumentation for the measurement of physiological functions (and sometimes for their control, enhancement, and even replacement, as in electric "pacemakers" for ailing hearts, hearing aids, and artificial kidneys) is a fertile part of the bioengineering field, but by no means all or even most of it. To begin with, instrumentation is not limited to monitoring life processes, but extends to such complex tools as high-power electron microscopes, radar-like probes to find tumors, laser beams for delicate eye operations, and hundreds of other devices used in research, diagnosis, therapy, and general clinical practice. Moreover, bioengineering contributions are not restricted to "hardware." Most likely even greater results will come from biomedical applications of such engineering methods as system theory, operations research, simulation, servomechanism theory—in short, not so much gadget design as engineering ways of thinking.

Here again the computer will come to play an important role. Already used in the collection and sifting of medical records, the electronic computer is beginning to be used to evaluate routine data such as blood tests and urinalysis and aid the physician in making a diagnosis by comparing his patient's symptoms with information stored in the computer memory.

Beyond that, and without indulging in science fiction, imaginative bioengineers are already speculating about technological approaches to such problems as blindness, uneven aging of bodily functions, unequal access to medical care and other social services, and the greatest human problem of all—the roots of aggressive behavior.[2]

TECHNOLOGY AND THE FINE ARTS *have influenced one another since antiquity*. Of the Seven Wonders of the World—the great pyramid at Gizeh, the wall and hanging gardens of Babylon, the statue of Zeus at Olympia, the temple of Diana at Ephesus, the tomb of Mausolus at Halicarnassus, the colossus of Rhodes, and the lighthouse of Pharos—most were works of art as well as great engineering achievements. Elaborate stage effects were a commonplace of the Greek theater during the Golden Age. The actor representing the god who appears at a play's critical point to resolve all difficulties of the plot—the *deus ex machina*—was actually hoisted into place by a machine—a contrivance of pulleys, wheels, levers, and other mechanical devices. Gadgets ranging from steam-operated temple doors to quite complex automata, used almost invariably for nonutilitarian purposes, proliferated throughout the Hellenistic period and beyond: among the gifts sent in 807 by Harun-al-Rashid to appease Charlemagne was an elaborate clock adorned with moving figures. Medieval and Renaissance artists were first of all superb artisans. The best-known example is Leonardo da Vinci (1452–1519), who not only painted the *Mona Lisa* and *The Last Supper* but was also the first to rationalize basic mechanical problems; his famous notebooks (some of which have only recently come to light) are filled with hundreds of engineering designs, including his valiant try at solving the problem of powered flight.

The Industrial Revolution created opportunities for further interactions between technology and the arts, notably through architecture. In the Great Exhibition of 1851, not only were the materials and products of a burgeoning industry shown side by side with the products of what were increasingly called the "fine arts," but the building housing the exhibits, the Crystal Palace (chapter 2), was itself a remarkable synthesis of technology and art. It was designed by the nonarchitect Joseph Paxton (1801–1865), was entirely of glass and iron, and marked the first time that iron had been used in architecture in an esthetically significant manner.[3]

The separation of the "fine arts" from contemporary life was deplored by William Morris (1834–1896), a superb designer who admired the Crystal Palace but hated the tastelessness of the industrial products displayed in it. Morris urged a return to the unselfconscious artistry of the craftsmen of an earlier day and himself led the way by turning his attention to such everyday artifacts as wallpaper, textiles, and furniture. His strictures made little impression on his fellow artists and even less on the industrialists of the day, although at least one of them, the German electrical manufacturer Emil Rathenau (1838–1915), did employ the architect Peter Behrens (1868–1940) as an industrial designer. It was a pupil of Behrens, Walter Gropius (1883–1969), who founded an enormously influential school of architecture and design at Weimar in 1919, the *Bauhaus*, in which "the principle of training the individual's natural capacities to grasp life as a whole, a single cosmic entity, should form the basis of instruction."[4] To this end, Gropius, his successor at the Bauhaus Ludwig Mies van der Rohe (1886–1969), and a little band of willful men in other lands—Frank Lloyd Wright (1869–1959) in America, Le Corbusier (1887–1965) in France, and Luigi Nervi in Italy—decided to do away with architectural traditions dating back to before the Industrial Revolution and to create a style that made full use not only of such newly available structural materials as steel, glass, and reinforced concrete but also of construction methods made possible by technical innovations such as mass production, prefabrication, and automation.[5]

Technology has influenced the fine arts as well as the applied. In the USA, it was a technological medium, photography, that provided the first new departure from established art norms in the work of Alfred Stieglitz (1864–1946); and the development of cinematography as an artistic medium likewise owes much to American film makers such as David Wark Griffith (1875–1948) and Charlie Chaplin. In painting, France led the way: the Cubist, Dadaist,

Futurist, Constructivist, Purist, and Surrealist movements were obviously influenced by the industrial environment, sometimes by way of reaction to it. Beginning in about 1960, a new movement has appeared (originally in New York, although spearheaded by European artists) that seeks to use the materials and techniques of the Second Industrial Revolution in creating works of art. The debut of this movement, which is characterized by the collaboration of artists and engineers, is usually taken to be the showing of the ephemeral sculpture "Homage to New York" by the Swiss artist Jean Tinguely and the Swedish-born engineer Billy Klüver. This work was a contraption made up mainly of junked parts of various machines and could move, at the same time producing assorted noises, visual effects, and smells, to the obvious delight of an invited audience. It was "programmed" to set fire to itself, begin to disintegrate, crawl to the pond in the Sculpture Garden of the Museum of Modern Art, and throw itself in, and though it did not quite make it to the water's edge (in fact, the debut turned into a fiasco), it had made its point: "*l'art éphèmère*," said Klüver, "creates a direct connection between the creative act of the artist and the receptive act of the audience, between the construction and the destruction; it forces us out of the inherited image and into contact with ever-changing reality."[6]

The idea of using automatic machines to animate works of art goes back, as we have seen, at least to the Middle Ages. Clockwork mechanisms dating back to before the Renaissance abound in Europe. Similarly, electric motors were used in 1931 in the earliest "motorized mobiles" of the engineer-sculptor Alexander Calder, which seem to have led to a profusion of moving objects by other artists—the so-called kinetic art. Electronic technology and synthetic materials have opened up new possibilities that are exciting to artists and engineers alike. An international organization, Experiments in Art and Technology, was founded in 1965 to "explore the possibility of a work which is not the preconception of either the engineer or the artist but is the result of the exploration of the human interaction between them."[7] The movement is not limited to the representational arts, and includes such art forms as the dance and music.

The impact of electronics on music has extended far beyond the generally acknowledged contributions to recording and reproduction. Steering away from mere juxtaposition of recorded natural sounds (*musique concrète*, chapter 3) and disregarding the Hammond Organ and other electronic instruments, composers led by Edgar Varèse (1883–1965), Vladimir Ussachevsky, and Karlheinz Stockhausen have sought to create new sounds from electronic oscillators, filters, reverberators, and other devices that electronics puts at their disposal. Elaborate instruments such as the "Moog synthesizer" invented by the Dutch engineer R. A. Moog are commercially available. Electronic music has been most successful in the theater, ballet, film, and television. It has had a lesser impact on the concert stage, possibly because no towering musical genius has so far interested himself in electronic composition, but some critics see it as the mainstream of music's future.

The most advanced joint efforts have sought to apply computers in the arts for visual, audio, and kinetic efforts. Punched cards, tapes, or sheets of figures are by no means the only "printouts" or final products of the computer: it can also produce graphic outputs —ephemerally on a cathode-ray-tube screen, or permanently as tracings, oscillographs, or other "hard" copy—and operate lights, move stage platforms, and produce sounds. Mechanical means of producing music have a long history. The barrel-and-pin music box dates back to the 18th century. The barrel organ and perforated-roll pneumatic instruments (of which the player piano, the calliope, and the merry-go-round orchestrion are more recent examples) are not much younger. In 1813, Beethoven composed *The Battle of Vittoria* (also known as the "Battle Symphony") for the panharmonium, a mechanical orchestra constructed by his friend Johann Nepomuk Mälzel (1772–1838), who later invented the clockwork metronome still used by musicians to fix tempi; and in the 20th century, ranking composers such as Paul Hindemith (1895–1963) and Igor Stravinsky (1882–1971) have written for the player piano, fascinated by the possibilities of piano music for an unlimited number of fingers.

All these techniques are based on recording what the composer has written. The electronic computer, which can vary volume, pitch, length, and sound quality through an infinity of gradations, is not only vastly more versatile than a player piano, but can, so to speak, participate in the composing process itself, by offering an instantaneous selection of note combinations programmed according to preconceived rules or systems of composition and even making the selection in a random way. The fascinating possibilities of this new technique are sure to attract the attention of major composers ultimately. Similarly, the possibilities of computers in producing visual patterns, geometric or abstract, are limitless.

Computers promise to play a major part in the art of the future. Whether that promise will be fulfilled depends on the caliber of the artists who elect to make use of this astonishing new medium.

TECHNOLOGY AND THE PEDAGOGIC ARTS *are the conjunction of advanced electronic systems such as television and computers and efforts aimed at the enhancement of learning at all levels.* Older devices, ranging from hornbook and blackboard to the widely available "audio-visual aids" of the precomputer age, are familiar: the phonograph and tape recorder, the slide and movie projector, the educational broadcast and television show. More advanced techniques let the student interact with the system. Audio tapes used for basic language instruction may be put under the control of the individual learner, who can rerun a section he did not get the first time or even record his own voice for comparison with the teacher's. Television systems have been set up for the most varied uses: playing back a prerecorded lecture-demonstration from a videotape cartridge; multiple viewing of an otherwise inaccessible event, such as a delicate experiment or operation; reproducing a greatly enlarged, live microscopic display for a big class; or relaying a lecture to audiences elsewhere. Interaction has been made possible in the last case, too: in one type of system, an unseen student may signal the lecturer and ask him a question over a radio link.

However, a note of caution has been sounded by some thoughtful observers, who fear that education may prove to be one of the fields in which the promise of technology has been oversold. The advent of the electronic computer has opened up fantastic possibilities for automatic, individualized instruction. According to this vision, a dazzling array of goodies beckons, if not from the shelf, at least from the drawing boards. Immense quantities of data can be stored and readily retrieved at millions of terminals—not only catalogs of library books but the contents of the books themselves, as well as pictures, audio and video tapes, films, and three-dimensional laser holograms. The retrieved information, heard or seen, transitory or more permanent hard copy, may take the form of instruction, highly individualized and adaptable to the student as to content, progression rate, methodology, and ultimate goals. Teachers freed from the routine, repetitive tasks that go with rote learning and its monitoring can devote themselves to the more humane task of catering to the needs of individual students. The students themselves need no longer be grouped in classes or grades arranged by age—each has what amounts to a private tutor. In a far-gone version of this utopia, the school is transformed into a testing, counseling, and guidance center, since—babysitting apart—computer-assisted instruction might just as well take place at home terminals; only learning activities that need special facilities or team efforts (laboratories, seminars, athletics, music) still require attendance at a central location.

An educator may surely take a jaundiced view of this rosy prospect without being thought irrational. A critical approach need not betoken mindless opposition to all incursions of technology into education, nor an attitude that would allow a teacher (or a physician, playwright, or bureaucrat) to appreciate the advantages of technology in every field of human endeavor except his own, thank you. Even Columbia University philosopher Charles Frankel—certainly no technocrat—finds some good in bringing technology to education. Once the limitations of technology are recognized, he sees definite advantages to acquainting students with it, if only because most will live out their lives in a world in which the technological style is very much present. Beyond that, he sees technology as having made an immense contribution to the moral and intellectual progress of mankind:

> The technological style simplifies and tidies our thinking processes, and, by introducing impersonal standards, brings to collective decision-making the coloration of objectivity and impartiality. . . . It takes the sheen off subjective preferences, family loyalties, class prejudices. Its breaks the crust of established institutions, exposes them in their inefficiency and superstition, and implicitly asks how they can be remade to serve practical human purposes better. And it imposes on people in different social stations and cultural climates something like a common set of values, habits, and ideas. In sum, technology is a form of culture, an instrument of education and of moral education. And it is such whether or not it is used deliberately in schools.[8]

But the limitations are there, and those who disregard them are victims of hubris, the prideful defiance of the gods by the doomed hero of Greek legend. Although technology can make things explicit, it is of little use where the greatest need is for the implicit: the need for personal attention, for immediate responsiveness, for models of behavior and achievement to emulate. Above all, educational technology cannot make up for such failings as low motivation, limited learning ability, and wrong or nonexistent goals—nor can it solve all the societal problems laid at education's door, from urban blight to racial prejudice.

That theme has been taken up with a vengeance by the Harvard computer expert A. G. Oettinger in a biting critique of educational technology's hard sell. In a book-length status report, *Run, Computer, Run,* he argues convincingly that in the foreseeable future, more technology will *not* lead to more efficient, individualized instruction, since any economically and technically feasible scheme of Computer-Assisted Instruction (CAI) presents such insurmountable difficulties of programming, scheduling, and monitoring as to make it highly unlikely for years to come for CAI to go much beyond simple drill-and-practice sessions of carefully limited subject matter in more or less synchronized sequences that approximate the "lock-step" deprecated by critics of present-day mass education.[9] The book serves as a critique of the fairly elaborate efforts that have been under way in the USA for some years to study the problem on a nationwide basis, sparked in part by the massive urge of the computer industry to conquer yet another market. A presidential commission, a committee of the National Academy of Engineering, a major foundation, and more than one professional society have been active in this direction.[10] Many of their findings support another contention of Oettinger's, that schools would have to be drastically reorganized to make them receptive to the massive reforms that would be needed. The "combination of institutional rigidity with infant technology preclude really significant progress in the next decade if significant progress is interpreted as widespread and meaningful adoption, integration, and use of technological devices within the schools."[11]

Where, then, does the greatest promise for an economically feasible application of technology to education lie? Not in substituting inert devices for teachers, nor in imagining that technology can serve as a panacea for all the ills of education from ignorance about the psychology of learning to poor management, nor yet in wasteful premature attempts to spread the tentative results obtained in a few moderately successful programs to as many primary and secondary schools as possible. Individualized instruction must remain a distant dream for a while longer. Instead, the greatest effort might more profitably go into substantial, prolonged, in-depth support of the most promising ideas, with costs of development and risks met by public as well as private funds. Quite likely higher education, with its greater resources available per student, is a more fertile ground for experimentation than primary or secondary education. At any rate, some of the best uses of computers in instruction yet achieved have come in college-level courses enhanced by such aids as computer-generated displays illustrating the dynamic (changing) results of applied mathematics and statistical phenomena; and CAI seems to be more profitably applied to professional instruction in medicine and engineering than to high-school subjects.

Finally, computers are not all that technology has to offer to education. The use of telecommunications, backed up by cheap and reusable audio and video magnetic tape, holds tremendous promise; so do many other devices. Above all, the goals of education must be set independently. Do we want to lift standards? select the most relevant among a glut of less useful information? expand the social functions of the educational system? Those are some of the long-range purposes identified by Frankel, who adds characteristically, "They require philosophy, not merely engineering." Oettinger agrees: "What *needs* to be done is a human problem," he concludes. "Technology only determines what *can* be done."

TECHNOLOGY AND HUMANISTIC STUDIES *are connected by one of the most recent bridges between the "two cultures."*[12] In the past, technology supplied few services to these studies beyond the production of printed books and similar scholarly apparatus. Here, as in so many other fields, new engineering tools and methods (for instance, computers and system theory) are beginning to have an impact on such diverse studies as linguistics, literary criticism, and historiography. An example of a field that reaps a particularly rich harvest of technological benefits is archeology.[13]

Excavation, still the foremost means of learning about man's prehistory, is being increasingly supplemented by aerial prospecting, novel earth-handling techniques, scientific dating methods, and other ways of putting physical instrumentation at the archeologist's disposal. In the West, two centers play leading roles: the Research Laboratory for Archeology in Oxford and the Museum Applied Science Center for Archeology in Philadelphia.

Aerial photography is itself an important aid in tracing archeological formations and identifying likely topographical locations not obvious from the ground or subsequently obscured by geological events and by the works of man (including agriculture). When combined with spectroscopic techniques by which separate images in colors ranging from infrared to ultraviolet are superposed, the mappings often reveal hidden features that point the way to a likely dig. Once a site is identified, buried artifacts and structures may be located near the surface by irregularities in patterns of electrical resistivity, and deeper down by magnetic measurements. Under particularly favorable conditions (soil that is slightly magnetic and very uniform), it is possible to detect differences between the magnetic properties of the soil and those of man-made structures as much as 6 meters deep. The ancient city of Sybaris in Italy was located and the street plan of another town, Elis in Greece, was mapped out by this method. Other innovations include underwater exploration by the entire range of novel techniques becoming available for it, from Self-Contained Underwater Breathing Apparatus (SCUBA) to two-man submarines and "side-looking" sonars that can map out substantial areas acoustically; and sifting and flotation techniques that increase the probability of separating small artifacts and plant and animal remains from soil.

Two scientific methods of dating archeological materials have had a tremendous impact. The first, carbon dating, is limited to organic (living) substances such as wood, leather, and fiber, and consists in measuring the radioactive decay of the carbon isotope ^{14}C that is always present in such materials.[14] This decay has been calibrated against that of substances whose age is known, such as tree rings and the independently dated artifacts of the various Egyptian dynasties.

The second method, thermoluminescence dating, applies to certain inorganic materials that contain no carbon, notably fired clays. Electrons at energy levels above normal are sometimes trapped in such clays and when a sample fragment is ground up and rapidly heated, these electrons are released and give off an amount of light proportional to its age.

The example of archeology shows that the applications of technology to humanistic studies thus need not be limited to "gadgetry," but can take their place alongside other methods available to scholars.

Monad emblem of Technocracy, an American social movement that flourished in the 1930s and advocated managerial organization of industrial society. The emblem was adapted from the Chinese yin-yang symbol "signifying unity, balance, growth, and dynamic functioning for the security of the life processes."

Chapter 6 Ideologies of Technology

TECHNOLOGY AND A SOCIAL ORDER BASED ON HUMANISM *are seen as opposites by some thinkers and as mutually reinforcing by others.* The modern concept of humanism is most frequently associated with the Renaissance, the return to classical ideals and forms, a concern for the highest human values—and what can they have in common with contemporary technology? A great deal, replies the technological determinist. It was the advent of modern technology that did away with slavery, raised the status of women and children, made social welfare a reality, and put freedom from hunger and insecurity within everyone's reach; it might—who knows?—some day do away with all the remaining barbarities, including war, once all men are free from want. This rationalist, materialistic view characterized most liberal and radical ideology of the 19th century, in which the idea of progress was almost indistinguishable from material progress; the voice of the one notable exception, the Danish philosopher Søren Kierkegaard (1813–1885), prophet of existentialism, remained virtually unheard by his contemporaries.

The social potentialities of the technological order seemed particularly great in the USA, if only it could preserve its traditionally open society—the first to begin to approach the Napoleonic ideal of *la carrière ouverte aux talents*, of social mobility, especially for the able. (Napoleon's own military career and imperial trappings tend to hide the fact that he was above all a superb administrator, far ahead of his time in recognizing that equality of opportunity was more than a revolutionary slogan—it was a necessity for any well-managed organization, if only to avoid the discontent of frustrated men of talent; every one of his soldiers, he said in a famed metaphor, carried a marshal's baton in his knapsack.)[1] Yet liberals looking around America at the end of the 19th century were hard put to see how the promise of technology was being fulfilled. The condition of the laboring masses, their ranks continually swelled by waves of immigrants, was parlous. The Civil War (1861–1865) had brought on an industrial development that took off in a laissez-faire atmosphere of almost no government regulation (or even income tax), creating huge private fortunes and a society that was far from egalitarian. It was a time of great political and economic strife, punctuated by recurring financial crises such as the Panic of 1873, which was followed by a drop in agricultural prices that ruined thousands of farmers and small proprietors who could not keep up payments on mortgages taken on when prices had been high.

The plea for a better, more humane, just, rationally ordered society built on a more equitable distribution of the new wealth, the mainspring of European socialism (chapter 1), was heard on the other side of the Atlantic with a strong American accent. Its loudest voice was that of Edward Bellamy (1850–1898), whose romantic novel *Looking Backward: 2000–1887* became a bestseller almost immediately after its publication in 1888. In the novel, the narrator wakes up mysteriously in the year 2000 and compares the society he finds with 19th-century America. Bellamy borrowed heavily from the socialists in describing the future—there is complete equality of income in his classless society and all means of production are in the hands of the state—but he avoided labeling his ideas as "socialist" for tactical reasons: the word, he wrote to a friend, "smells to the average American of petroleum, suggests the red flag, all manner of sexual novelties, and an abusive tone about God and religion." (One is reminded of the reply made by the unabashed Louisiana demagogue, Senator Huey Long, shortly before his assassination in 1935, when he was asked if the USA might ever embrace fascism: "Yes," said the Kingfish, "but *we* shall call it antifascism!")

But Bellamy's vision differed from Marx's in some important ways. To begin with, the new order comes about without a class struggle. The state, far from withering away as Marx hoped, is a strong, highly centralized institution without any soft ideas about democracy; political parties are abolished and the government is run by state-appointed managers, while the rest of the working force is organized into a disciplined Industrial Army, whose generals —one for each "trade" or guild—are elected by the retired members of the guild (those over 45, who alone have the vote) and appoint all the lower ranks.

Bellamy's utopia stands out among other examples of the genre for two reasons. The first is that the novel had real, immediate political consequences. Bellamy clubs sprang up all over the USA; during the brief heyday of the Populist party (whose candidate polled over a million votes in the 1892 presidential election), much of its support came from the adherents of Bellamy; and some of his ideas ultimately found their way into the platform of the two major parties, which as always dealt with the threat of a third party by adopting some of its planks. But the prophecy is remarkable for a second reason: his prediction of a new sort of society, not exactly socialist and certainly not capitalist—the "managerial" society—the advocacy of which runs like a red thread through American intellectual history; the concept is given relevance by the considered view of many observers that it is precisely the sort of society that, for better or worse, prevails today.

Outstanding among the advocates was Thorstein Bunde Veblen (1857–1929), who grew up in Norwegian farm communities in the Middle West, heartland of agrarian reform and cradle of another abortive third-party movement (the Progressive party, which polled over 5 million votes in the 1924 presidential election). Veblen was associated with first-class universities both as student and teacher, but his background set him apart from the dominant American society of his day, which he viewed with growing skepticism. He was particularly scathing about the "lag, leak, and friction" typical of an industrial system run on behalf of vested interests whose beneficiaries, the "kept classes," thought more about their own luxurious "conspicuous consumption" than about the common weal. A society reorganized on sound engineering principles, run by professional managers, was the proper way. In 1918 Veblen ended his formal academic career (he last taught at the University of Missouri) and came to New York City; after a year, he joined the faculty of the just-founded New School for Social Research, where he lectured on subjects such as "the social functions of the engineer" and "productive use of resources" and wrote a series of articles that were published in 1921 as *The Engineers and the Price System.*[2]

In this slim volume, Veblen said that technology was "impersonal and dispassionate, and its end very simply to serve human needs, without fear or favor or respect of persons, prerogatives, or politics." He called on the engineers and technicians, "the General Staff of the industrial system," to become the vanguard of a revolutionary movement that would bring about a nonviolent overturn, without social disorganization; put an end to absentee ownership; and reorganize industry on a more efficient basis. He specified no underlying political structure. Society would become a self-administered industrial organization in which there was no need for politicians, whom he considered below contempt: "A degree of arrested spiritual and mental development," he remarked in a later essay, "is, in practical effect, no bar against entrance into public office." At the same time, he had his doubts about how soon the changes he foresaw would come. American industry could, he conceded, muddle through for a bit longer, inefficiencies and all; the engineers were in any case on the side of the owners; labor unions only wanted more for their members, not a thoroughgoing change; and the general population seemed to be largely satisfied with the existing system.

These reservations accord with ideas that Veblen had been formulating ever since he had published his first and best-known work, *The Theory of the Leisure Class,* in 1899. He understood, better than most of his radical contemporaries, how decisively the business civilization had carried the day. He saw conflicts between business and industry as symptoms of a "culture lag," which arose because institutions and organizations failed to keep pace with social changes stemming from scientific and technical advances.

Veblen's concept of the culture lag was taken up by several sociologists, notably William Fielding Ogburn (1886–1959), whose *Social Change, with Respect to Culture and Original Nature* (1922) initiated a whole school of thought. Ogburn's thesis that social change is primarily the result of socio-cultural changes, notably invention (rather than mainly geographical, biological, or ideological factors), is probably the most widely accepted theory among present-day Western social scientists.[3] But Veblen's ideas had a further effect in creating an entire social movement, technocracy, which held the stage in America for a brief dozen years. It was contemporaneous with the Nazi movement, of which it was uncomfortably reminiscent in a number of ways, though not in popularity.

TECHNOCRACY *was a social movement that sought to put a system of optimum industrial production and distribution controlled by technicians in place of the capitalist price system.*[4] (The word has since acquired a broader meaning: Webster defines it as the "management of society by technical experts.") The movement was the brainchild of Howard Scott, one of a group of enthusiasts originally centered around the New School for Social Research and strongly influenced by Veblen. During the 1920s, Scott had also had contact with the syndicalist IWW (the International Workers of the World, or "Wobblies"), the most colorful American radical organization of the day, which advocated direct action and stood between socialism and communism (the "2½" International). In 1932, Scott found support for his cause from a group at Columbia University, then as now creditably hospitable to off-beat ideas; although the alliance lasted only briefly, it lent a measure of respectability to the movement, which now established itself as Technocracy, Inc.

Technocracy, Inc. (the organization still exists, mainly in the western USA) had a negligible impact on American life; what little force it could muster was dissipated by poor organization and totalitarian trappings rejected by most Americans who have outgrown the Boy Scouts. (The movement was not run democratically; it had a distinctive emblem, the Monad symbol, a circle divided into two fields by an S-shaped curve; the members wore gray shirts and suits and often even painted their cars gray.) But the managerial ideology turned up again in the New Deal of President Franklin Delano Roosevelt (1882–1945) and thus indirectly achieved considerable public influence, even though technocracy itself remained a minor movement. Was that because its goals were too boldly stated? A political commentator reviewing the case in 1940 thought so. "Technocracy's failure to gain a wide response," wrote James Burnham, "can be attributed in part to the too-plain and open way in which it expresses the perspective of managerial society. In spite of its failure to distinguish between engineers and managers (not all engineers are managers—many are mere hired hands—and not all managers are engineers) yet the society about which the Technocrats write is quite obviously the managerial society, and within it their 'Technocrats' are quite obviously the managerial ruling class. The theory is not dressed up enough for major ideological purposes. It fails also in refusing to devote sufficient attention to the problem of power, which so prominently occupies communism and fascism."[5]

THE MANAGERIAL REVOLUTION, *by James Burnham (1941),* from which the foregoing quotation is taken, is another formulation of the thesis that the society of the future is likely to be neither capitalist nor socialist, but a distinct new form, toward which widely disparate societies of 1940—the Soviet Union, Nazi Germany, and New Deal America—were tending. Burnham was less successful in forecasting the actual shape of things to come: as the socialist writer George Orwell (1903–1950) pointed out in a biting postwar essay, Burnham predicted a German victory over Britain and the Soviet Union when the Germans were winning—and later a Russo-Japanese alliance when the Japanese were in the ascendant and a Soviet world conquest while their troops were sweeping all before them—and he seemed to show a shocking worship of power in general and an open admiration for the methods of Hitler and Stalin in particular. In Orwell's summary, this is Burnham's thesis:

> Capitalism is disappearing, but Socialism is not replacing it. What is now arising is a new kind of planned, centralized society which will be neither capitalist nor, in any accepted sense of the word, democratic. The rulers of this new society will be the people who effectively control the means of production: that is, business executives, technicians, bureaucrats, and soldiers, lumped together by Burnham under the name of "managers." These people will eliminate the old capitalist class, crush the working class, and so organize society that all power and economic privilege remain in their own hands. Private property rights will be abolished, but common ownership will not be established. The new "managerial" societies will not consist of a patchwork of small, independent states, but of great super-states grouped round the main industrial centers in Europe, Asia, and America. These super-states will fight among themselves for possession of the remaining uncaptured portions of the earth, but will probably be unable to conquer one another completely. Internally, each society will be hierarchical, with an aristocracy of talent at the top and a mass of semi-slaves at the bottom.[6]

Writing six years and a world war later, Orwell argued that the thesis was untenable, not because Burnham's view of contemporary trends was false (most of the developments of the first half of the 20th century admittedly did point towards oligarchy), but because Burnham assumed that the trends were irresistible. They were not: owing partly to technology, the dominion of man over man is ceasing to be necessary and class distinctions based on a need for drudgery are sure to disappear—though they might turn up in a new form, for deep-seated psychological reasons that have nothing to do with external necessities.

We shall presently return to the question of the implications of the new industrial order for human values. For the moment, we note that all the views we have examined—Bellamy, Veblen, technocracy, Burnham—antedate the Second Industrial Revolution. Surely that event has had a profound effect on society, at the very least on the political and economic structures of highly industrialized nations? Not a bit of it, to judge from the writings of most economists. The great innovation of the first half of the 20th century was the result of the work of Lord Keynes (1883–1946), whose prescriptions for the enlightened public control of the economic life of nations (by adjustment of the interest rate and of the amount of public spending, and by other government measures) are said to have saved capitalism.[7] The paradox that the free-enterprise system can carry on only under close government control bothered capitalists when they first saw it in large-scale action in the 1930s, but they have grown accustomed to this phase and have seen society move a long way beyond it. What does this society look like and where does it seem to be heading under the impact of the Second Industrial Revolution? A determined attempt to answer these questions has been made by J. K. Galbraith, a Harvard economist with extensive experience in public life. This iconoclast professes to see, a quarter of a century after Burnham, the managerial revolution in full swing, though not exactly along the lines he foresaw.

THE NEW INDUSTRIAL STATE, *by John Kenneth Galbraith (1967)*, is a systematic mapping of our industrial society, ranging far beyond purely economic aspects and showing how it differs from earlier versions, especially in its consequences on social and political behavior. In this work, Galbraith used a technique perfected in an earlier bestseller, *The Affluent Society* (1958), of forcing his fellow economists to pay attention to him by appealing over their heads to a larger public than ordinarily follows economic disputation. This procedure has not endeared him to his colleagues, many of whom criticize his ideas as neither sound nor novel.[8]

Galbraith's industrial system is characterized by the presence of many large corporations, anonymously managed; large state participation and regulation of income; no major business cycles; much market research and advertising; a decline in the influence of trade unions; massive higher education; and a heavy interdependence of all these features. Dependence on the market, the classical law of demand and supply, has been largely replaced by planning. The corporation itself is the planning instrument; supply and demand are replaced by savings and the uses to which savings are put.

The "mature" corporation differs from the earlier "entrepreneurial" firm mainly in the way it is managed, by managers who are not responsive to the owners (stockholders), so that control has largely passed to the professional managers and the specialists on whom they depend (the "technostructure"). Decisions are made deep within this technostructure; top management really only ratifies the decisions and seldom overrules them—it would need a second lot of specialists just to gather the necessary information. Thus power has passed first from the land to capital, then to labor, and finally to organizations of specialists.

One consequence of this shift is that economic theory needs to be revised. The assumption that the corporation's overriding motive is to maximize profits can no longer be made: salaried managements are motivated neither by compulsion (as slave societies based on land values were) nor by compensation (as was typical of entrepreneurial capitalism) but by identification with the goals of the organization and by the desire for adaptation of the organization's goals to those of the individuals concerned. Opportunity to shape the organization's goals might be expected to motivate those with an urge to power.

Galbraith examines the goals of the technostructure: a secure earnings level, growth, technical virtuosity, and rising dividends; maximizing returns is not necessarily one of them. Also needed for success are (1) the management of specific demand, as by advertising, especially on television; and (2) regulation of aggregate demand (following Keynes) by which private and corporate income is regulated through taxation and government spending and planning. But employment also needs to be controlled. Growing mechanization has reduced the need for blue-collar workers and hence the power of the unions; yet unemployment of the uneducated coincides with shortages of the specially qualified, leading to a new kind of class struggle, between the educated and the uneducated. (In the USA, the struggle is intensified by the circumstance that many of the uneducated are also members of minority groups.) Low demand of labor leads to unemployment, high demand to the wage-price spiral and inflation. There is hence need for government intervention, first introduced systematically in the USA during the administration of John Fitzgerald Kennedy (1917–1963): the firm sets the minimum price and regulates specific demand, while the state regulates aggregate demand and sets maximum prices and wages. This regulation weakens the unions even further.

Galbraith sees the mature corporation as an arm of the state, and the state as an instrument of the industrial system. The technostructure does not bribe or buy politicians—it lacks the means (and will) and is in any case not obliged to do it since the state accommodates it well enough. The industrial state and the technostructure have common goals: stability, growth, education, and technical and scientific advance.

Galbraith agrees with other observers that the polarization of the world into superpowers and the resulting tensions have been useful to the technostructure. He cites, not without relish, the warning sounded by President Dwight David Eisenhower (1890–1969) just before he left office in 1960 that the USA should "guard against the acquisition of unwarranted influence, whether sought or unsought, by the military-industrial complex."[9] Yet defense need not be the mainstay of the industrial state—anything else will do as long as it is on a large scale: space, scientific research, communications, ocean exploitation, or climate management. Coexistence with the Soviet-type system is quite feasible, especially since it begins to converge with the industrial system in the capitalist countries. Successful socialist enterprises are likewise controlled by technostructures rather than by society. (The Soviet "firm" needs a trifle less autonomy than its capitalist counterpart, since the state does much of the planning for it; but it is not true that the Soviet economy is beginning to be dominated by the market—it is merely being somewhat more decentralized, that is, some of the planning functions are being shifted from the state to the firm.) There are other similarities: both the Soviet-type and the capitalist industrial system need large-scale production, elaborate organization, management of demand, autonomy over planning, regulation of total demand, and education. The USA tends toward the "seminationalized" firm, a corporation that is in effect part of the government; one may even speak of the "socialization" of the mature corporation.[10]

But the state must intervene also on behalf of social services, which are unimportant to the industrial system, and of esthetic considerations, which Galbraith sees as actually inimical to it, since they are highly individualistic and produce other constraints. To be sure, making the state the custodian of esthetics is less than a perfect arrangement, but there are large areas in which planning nevertheless must be done. Local transit, public housing, conservation, recreation, and forestry are all examples of areas that cannot rely on the market but for which significant planning is lacking.

Galbraith's analysis thus seems to bear out the earlier heralds of the managerial society—for that is what we seem to have got. Not that this state has come about by the conscious, deliberate efforts of power-crazed technocrats. No—what has happened suggests an even darker possibility: that technology has, so to speak, a life of its own that causes it, like some fairly elaborate monstrous organism, to batten on whatever is in its way and to assume forms that are not only inhuman but subjugate all human purposes to its own requirements. Moreover, this trend is seen as independent of the sort of political system that prevails.[11]

The idea that something has gone wrong with civilization is not new. The war of 1914–1918, with its senseless millions of dead, might have been expected to shake the cheeriest optimist's faith in progress. Western civilization was finished, on its way out—it would never again reach the pinnacle represented by, say, Beethoven's late string quartets—those were views that brought worldwide recognition to the German historian Oswald Spengler (*The Decline of the West,* 1918).[12] The French humanist Julien Benda (1867–1956) accused the *clercs* (writers and thinkers) of his day of betrayal, of thinking more about saving their skins than about leading mankind toward liberty and justice (*The Treason of the Intellectuals,* 1927). The Austrian psychiatrist and founder of psychoanalysis, Sigmund Freud (1856–1939), saw the suppression of man's natural urges by civilization as society's greatest malaise in *Civilization and Its Discontents* (1930).

The war of 1939–1945 did nothing to restore the intellectuals' faith in technological civilization. Writing as a silent opponent of the dying Nazi regime—and at the same time cowering helplessly at the receiving end of Allied bombs—the German poet and philosopher Friedrich Georg Jünger[13] concluded, in *The Failure of Technology* (1946), that technology was above all a demonic force that could be contained only if its growth was deliberately cut off before it crowded out humanity and destroyed all civilization—including technology itself. (The book's original title, *Die Perfektion der Technik,* refers to this ultimate apotheosis of technology.)

All these predictions of the impending doom were made before the Second Industrial Revolution. The hopeful respite provided by the defeat of large-scale fascism and the founding of the United Nations was very brief. With the advent of nuclear weapons, electronic computers, automation, the cold war, and the beginnings of the managerial society, the prophets of an imminent *Götterdämmerung* have redoubled their efforts. The warnings they sound deserve our attention in proportion to the extent to which they are logically derived from facts and visible trends. The most serious analysis of the unwelcome direction the technological society is taking has come from Jacques Ellul, a brilliant French scholar with a background in law, history, and the social sciences.[14]

THE TECHNOLOGICAL SOCIETY, *by Jacques Ellul (1954)*, is by far the most far-ranging and systematic account of the threat that technology poses and of the price that its benefits exact. (In a later contribution, the author enlarges on the invariable ambivalence of technical advances by showing that they raise more problems than they solve, that the pernicious effects are inseparable from favorable effects, and that every technique implies unforeseeable effects.)[15] The original title, *La technique: L'Enjeu du siècle* (Paris: Librairie Armand Colin, 1954), which means "Technique: The Stake of the Century," comes nearer describing that viewpoint. Here we meet a semantic difficulty. The French word *technique*, and the German *Technik*, are usually put into English as "technology" (except by Lewis Mumford, who calls it "technics"), and the English word "technique" means something else: "a technical method of accomplishing a desired aim" (Webster). Ellul's usage of *technique* in both French and English is vastly broader still: the totality of all rational, efficient methods in every field of human activity. In other words (those of his translator, John Wilkinson), technique is "nothing less than the organized ensemble of *all* individual techniques which have been used to secure any end whatsoever." This omnibus definition extends the concept of technique far beyond its ancestor, machine technology (which is "pure technique"), to embrace all social, economic, and administrative life, including for instance such particular techniques as education, the judiciary, the several social sciences, organized sport, propaganda, and in fact all that the historian Arnold Toynbee calls "organization" and Burnham calls "managerial" action.

In this view, technique is not only the means to all sort of ends; it has ominously become an end in itself, pushing us inexorably toward loss of personal liberty and the disappearance of human ends. This apocalyptic outcome could be perhaps avoided by the concerted action of a large number of people willing to assert their freedom by upsetting the course of this evolution, but Ellul plainly does not place much faith in our ability to do so. He is particularly scornful of proposals to do it by technical means, which would only compound the dehumanizing process. (He does allow, without irony, that his forecast could also be invalidated by a nuclear holocaust or by divine intervention.)

In a brief historical review, Ellul traces the distaste of the ancient Greeks for technique less to the traditional explanation (that slavery made improvement of practical life unnecessary) than to a deliberate rejection of what they recognized as a potential monstrosity. The Romans perfected both civil and military social technique, mainly through law, aimed at strengthening the internal coherence of society. They were less successful in material techniques, except during the two centuries from 100 B.C. to A.D. 100. In that respect, both Greece and Rome relied on the East, a fact that is contrary to the stereotypes of the Buddhist and Moslem East as passive, fatalist, and contemptuous of life and action, and of the Christian West as active, conquering, and turning nature to profit. In the Dark Ages, Europe went on depending on the East for technical innovation; besides, Christianity often opposed such innovation on moral grounds. Likewise, Ellul sees the Renaissance as a period of relatively few technical advances and of preoccupation with humanism, the supremacy of man over means.

It was the Industrial Revolution that created the modern state, with its strong dependence on technique and on systematization, unification, and clarification in all fields. Technique blossomed in the 18th and 19th centuries because of the conjunction of five phenomena: the fruition of a long technical experience, going back to A.D. 1000; the expansion of population (and hence of markets and of human material); a suitable economic milieu, stable yet capable of change; the new fluidity of the social milieu; and a clear technical intention on the part of important groups—the state, the capitalists, and the bourgeoisie. Everybody seemed to be *for* technique, except perhaps the peasants, who were slowly ruined.

In characterizing modern technique, Ellul sees little likeness to the technique of earlier times, when it was limited in breadth, technical means, and extent, and when there was a choice between an active and a passive society. In our day, there is no choice; technical progress is irreversible and feeds on itself; technique has become independent of moral judgments and makes its own morality, which is that all that is possible is necessary; technique is universal and autonomous—it conditions evolution in the economic and social spheres rather than being conditioned by them. Once these characteristics of the monster have been established—Ellul likens them to "the psychology of the tyrant" (that is, of technique)—we are taken on a devastating tour of his digestive tract (the economy), the circulatory system (the state), and the cellular tissue (man).

As to *economy*, Ellul agrees with both Marx and Keynes that technical progress is indispensable to it. Such progress inevitably calls for greater concentration of capital and thus leads to either a state or a corporate economy. The state intervenes in any case to impose planning and other controls; economics, at first a technique of observation, becomes a technique of action as it goes from gathering statistics, accounting, modeling, and public-opinion research to establishing norms and to planning, new tasks that imply total scope, centralization, coercion, and state control. Thus technique, once accepted, inexorably demands subjection to its laws; it is inherently antidemocratic and leads to elitism, standardization, conflict between personal choice and the state's requirements, between the natural and the artificial. What classical theory called "economic man" has been forced into being by technique.

The *state* is affected because it tends to organize all aspects of life, not only by well-established military, financial, judicial, administrative, and political techniques, but also in newer areas: transport, education, social welfare, communications, and in fact all the expensive fields (big science,[16] industrial subsidies, urban planning). The results are perhaps unintended: a progressive refinement of the state techniques, old and new; the replacement of politicians by technicians and the transformation of the nation into an object of the technical state, in which technique can proliferate unchecked by morality, public opinion, or social structure—in fact, independent of man himself; the subordination of the state's own structures and organs to techniques, for instance in law, which becomes largely an administrative proceeding, with "order and security" replacing justice as the basic tenet; the devaluation of political doctrines into mere justifiers for pragmatic action; and the ultimate transformation of the state into a totalitarian one—not necessarily brutal, but inhuman just the same. Ellul studies this specific phenomenon in some detail in another book, *L'Illusion politique* (1958).

Ellul's indictment of the effects of technique on man is particularly mordant. Modern life increases human tensions, changes the human milieu for the worse, modifies the concepts of time and motion, and creates a mass society. Ellul gives short shrift to proposals that would use technique to improve man himself, combat slavery, humanize techniques, reconstitute the unity of the human being, or otherwise enable man to intervene in his own evolution. He dismisses the social sciences involved in such attempts—they are nothing but more techniques, and one cannot create humanism by increased use of techniques. Education aims toward creating social conformism; work techniques have maximum productivity, not humanism, as their goal; vocational guidance, a totalitarian takeover of the young, only creates organization men.[17] The union of technological and psychological techniques has produced one technique to which Ellul had previously devoted an entire book, *Propagandes* (1962). Propaganda has made its way from commerce into politics: masses can be manipulated on both the conscious and the subconscious levels as the critical faculty is suppressed, spiritual development is arrested, and an unreal "verbal universe" is created. He cites several examples. Amusements, fully adapted to a technical society, only produce alienation and flight from reality. In sports, organization parallels industrial organization. Even medicine becomes a technique exploited by the state, for instance in criminology. Techniques encircle man—not any one technique, but the spontaneous convergence of all techniques, including the "human techniques" (the social sciences). The more that is learned about man, the less human he becomes, the more he resembles *l'homme-machine* of the atheist materialist Julien Offray de la Mettrie (1709–1751), who held that the soul must be an organic part of the brain and the nervous system. Far from integrating man, technique disassociates him, for instance, in the separation of thought and action during work on an assembly line. Every human technique appeals to the unconscious, including art. Moreover, human techniques necessarily aim to adapt man to a mass society—advertising is just one form of psychological collectivism. To the psychotechnicians, "reintegrating" the human personality into the mass order is a major aim; they are mistaken to think that they are thereby liberating man. Even spiritual activity is technically mediated; it is next to impossible to write a book, enjoy nature, or engage in political activity outside the technical framework.

Ellul is not impressed by the claims of technology to have improved man's lot. That is a hollow victory, he says, "gained at the price of an even greater subjection to the forces of the artificial necessity of the technical society that has come to dominate our lives." He ends on a characteristically somber note. Our efforts to rediscover a purpose or goal for human society are doomed; on the one hand intuition and ethical judgments are subjective and thus unequal to the task, and on the other hand the use of technical means to determine the ends of the technical society can be no panacea and is clearly self-defeating. Yet the book is subtitled, *The*

Stake of the Century, which would seem to indicate that Ellul does not consider the contest to be lost—but that it can be won by man only if he is aware of its grave import.

The reader unfamiliar with the methods of doctrinaire disputation should be forewarned about the value and utility of all these absolutes. It is not usually part of an established church's teachings that its articles of faith are only provisionally true; a system of political ideology seldom admits the merits of opposite views; a new theory in the social sciences (say, that all human behavior is conditioned by two opposing drives, the sexual and the death instincts) is not likely to point out its own deficiencies. When persuasion is the name of the game, we can expect neither a St. Thomas Aquinas nor a Karl Marx nor a Sigmund Freud to go about it by halves, timidly putting forward a set of tentative ideas and adducing their weaknesses.

It is in this context that the work of Ellul should be seen. When he ends his dissertation by suggesting that one should not really expect technicians or scientists to seek humane motives for the golden age they are creating, that "all that must be the work of some miserable intellectual who balks at technological progress" (that is, Ellul)—is he not really telling us that his prognosis will be right *unless* we do something, and that we must do something fairly quickly? He does not tell us what it is we must do: he attempts a diagnosis, not a treatment. (In the preface to the American edition, he mentions the solutions proposed by Emmanuel Mournier, Pierre Teilhard de Chardin, Ragnor Frisch, Jean Fourastié, and Georges Friedmann, all of which he airily dismisses as fanciful and unrealistic.)[18] In that respect he resembles the writers of the 20th-century antiutopias, whose influence was the greater because at least some of them were writers of the first rank, so that their ideas reached a large public.

THE QUEST FOR UTOPIA, *a search for the ideal society,* has a long history, going back beyond the *Utopia* (Greek for "nowhere") of that man for all seasons, Sir Thomas More (1478–1535), to St. Augustine and Plato. (We spoke of another utopian visionary, Sir Francis Bacon, in chapter 1; and Rabelais and Swift also contributed to the genre.) The utopian literature occasioned by the Industrial Revolution gave way to largely political programs in the 19th century, but toward its end the art form was revived by such writers as Bellamy and his English contemporary, the esthete William

Morris (*News from Nowhere*, 1891). Yet a new note, antiutopianism, crept in as early as 1872.[19] That year saw the publication of Samuel Butler's *Erewhon* (which is "nowhere" more or less spelled backward), a biting satire on the social and economic conditions of Victorian England, in which fear of technology is taken to its logical conclusion when the citizens determine to destroy all machines.[20] *Erewhon* finds an echo in the various 20th-century expressions of antiutopian sentiment, including opposition to subjugation by a mindless bureaucracy or by some other technical elite, opposition to violence and coercion for utopian goals, and finally opposition to all utopian ideals. These expressions are to be found not only in philosophic works ranging from the strongly individualistic views of Friedrich Wilhelm Nietzsche (1844–1900) to the more recent "incrementalism" of Karl Popper (*The Open Society and Its Enemies,* 1950, 1963) but also in *belles lettres*: Franz Kafka,[21] E. M. Forster,[22] Eugene Zamyatin, Aldous Huxley, and George Orwell.[23] We shall take a brief look at the last two, who have been probably the most influential.

Aldous Leonard Huxley (1894–1963), grandson of the great Victorian biologist and educator Thomas Henry Huxley (1825–1899) and brother of the zoologist Julian Huxley, had already made his name with a succession of witty novels describing the decadent society of postwar Britain when he published the most brilliant antiutopia of them all, *Brave New World* (1932). The novel describes the antiseptic, scientifically organized, termite-like world of the 7th century A.F. (after Ford—that is, Henry Ford, who has become the Deity of the new age), run as a tight caste system in which free will has been abolished by methodical conditioning (including hypnopedia, or sleep teaching) and servitude has been made acceptable by unlimited doses of chemically induced happiness, unrestricted access to the opposite sex, and other freely available distractions. Eugenics —in the form of the scientific control of breeding, including a process by which ninety-six identical twins can be reared by "ectogenesis" (outside the mother's body) from a single ovum—is one of the state's main concerns. The resulting population is permanently divided according to intelligence into Alphas, Betas, Gammas, Deltas, and the near-moronic Epsilons, all conditioned to be wholly unaware that their lot is not a happy one. One-third of the population remains on the land, and production is scheduled so that there is seven and a half hours of work for all, regardless of efficiency. This painless utopia is governed by a worldwide oligarchy of a few assorted experts, each an Alpha Plus no more recognizably human than the herd over which he lords it with such supreme assurance.

The theme that prosperity and unending happiness may be only perpetuated at the price of an inhuman sacrifice of all deep emotions, the suppression of individual freedom, and the division of society into leaders and led is not original with Huxley. The societies ruled by Plato's philosopher-king (in *The Republic*) or by Machiavelli's *Prince* are far from democratic. In the microcosm of Shakespeare's *Tempest* we find Miranda (innocence) and Caliban (ignorance) in thrall to Prospero (magic). Dostoievsky's Shigalov (in *The Possessed*) and Grand Inquisitor (in *The Brothers Karamazov*) are telling caricatures of utopian idealists. But Huxley's underlying idea, also put forth by the exiled Russian religious philosopher Nicholas Berdyaev (1874–1948), was that the problem was not how to achieve utopia but how to forestall it; that mankind had arrived at an age in which, Berdyaev's words, "cultured and intelligent people will dream of ways to avoid ideal states and to get back to a society that is less 'perfect' and more free."[24]

It is a short step from that to imagining a wholly malevolent society, a negative utopia in which all utopian goals are perverted and everything aims only at keeping the elite in power. In 1948, the English socialist George Orwell (whose real name was Eric Blair) looked at a world that had only just taken care of Hitler but had yet to deal with Stalin, and imagined what it might become under these circumstances in the three dozen years that remained before the last two digits of the date were reversed—before 1984.

In Orwell's *Nineteen Eighty-Four* a terrorized population, conditioned physically and mentally without letup and held in check by an ever-present Thought Police, unquestioningly follows the leader, Big Brother, no matter how often he changes his mind. There is a permanent state of war, now against one adversary, now with his help against another. Everyone must practice the art of "doublethink," of holding two contradictory beliefs in one's mind at once and accepting both of them. No individual thought—let alone action—is possible; the slightest attempt at revolt is swiftly suppressed and punished. "If you want a picture of the future," one of the elite tells the hero, "imagine a boot stamping on a human face —forever."

Obviously neither Huxley nor Orwell endorsed the nightmare societies they described (no more than does Ellul), quite the opposite: their books were meant to serve as warnings of what may happen if we do not watch out. The warnings have been heard, at least in the West. (Orwell's book has of course never been published in the USSR or in China, though copies of it are known to circulate clandestinely throughout eastern Europe.) Many of Orwell's locutions have entered the language, including "Big Brother Is Watching You" and, from his 1946 parable of Stalinism, *Animal Farm,* "All animals are equal but some animals are more equal than others." Orwell died a year after his book was published and never saw whether the world had moved nearer to the horrors he foresaw. Huxley, who outlived him by over fifteen years, got another crack at the predicting business. Ten years after Orwell's book, Huxley published *Brave New World Revisited* (New York: Harper & Row, 1958), in which he argued that the odds were more in favor of something like his forecast than Orwell's. He also pointed out that what with nuclear explosives, overpopulation, tranquilizers, brainwashing, and subliminal projection, his prophecies of twenty-seven years earlier were coming true (and were even being exceeded) much sooner than he had thought they would. Yet if we can trust the insider's portrait of the communist power elite drawn by the Yugoslav writer Milovan Djilas in *The New Class* (1957), we find it hard to escape the feeling that a despotic bureaucracy has grown there whose self-serving ends (and means) make for a society that leans more heavily toward *Nineteen Eighty-Four* than toward *Brave New World.*[25]

It would be fascinating to know what Orwell or Huxley might have thought of Ellul's *Technological Society.* Orwell died before the book appeared, but Huxley knew of it and had in fact recommended it to the Center for the Study of Democratic Institutions of the Ford Foundation-sponsored Fund for the Republic, Inc. In 1962, a conference on "The Technological Order" was convoked at the center (in Santa Barbara, California) by the editors of the Encyclopaedia Britannica; Huxley, who spent the last years of his life in California, attended. Ellul sent a paper in which he recapitulated the essence of his book, notably the contention that the artificial, autonomous technical milieu exists independently of all human intervention, that it is "self-determining in a closed circle." This image of technique forming an unbroken loop that adapts man to itself (rather than the other way around) caught the eye of more than one participant, including Huxley. The report of the discussion contains a tantalizing glimpse—the only one in print—of what Huxley thought about Ellul.

Mr. Huxley commented that technological development tends always to obey the laws of its own logic, but pointed out that the same thing is also true of, for example, art. This means, he said, that the question to be asked is whether the laws required by technology's own logic are the same as those required by human logic and he suggested that when looked at in this way there does indeed seem to be a good deal to be said for the theory of the closed circle.[26]

Another conference participant was Anatoli Zvorykin, top Soviet sociologist and historian of technology, who presented a major paper on "Technology and the Laws of Its Development"; it can serve as a convenient introduction of a view of technology that is not very familiar to most Western students.[27]

MARXIST VIEWS OF TECHNOLOGY *are developed in more or less strict accordance with the official philosophy, dialectical materialism,* by the method of looking for internal contradictions and in that way discovering the governing laws. Dialectical materialism was developed by Marx and Engels (and elaborated by Lenin) from the dialectic, a logical method used by the great German philosopher Georg Wilhelm Friedrich Hegel (1770–1831). In a simplified form of this logic, each concept (thesis) creates another (antithesis) and their interaction leads to a new concept (synthesis), which in turn becomes the thesis in a new triad. Dialectical materialism holds that men's material circumstances determine their views; or stated simply, that everything is material and all growth, change, and development take place through the struggle of opposing elements. The corresponding view of history and sociology ("historical materialism") sees class struggle as the only dynamic of history, which follows inexorable laws that are not influenced by individuals. Following this system of thought Marx predicted that the revolution would first occur in a highly industrialized nation, that placing all means of production in the hands of the state would be followed by greatly increased production and the creation of a classless society, that the state would then wither away, and that in the period leading to the revolution, the condition of the workers in bourgeois societies would steadily worsen. Anti-Marxist writers point out that as none of these things has in fact happened, the entire theory has been discredited.[28]

But according to orthodox Soviet interpretation, no other view is admissible: "Marxist-Leninist science gives the only correct answer to the question of the significance of technology for society." Using standard communist jargon (for instance, "means of labor" for tool, "object of labor" for raw material, and "technical means" for instrument or machine tool) and solemnly quoting from Marx, Engels, Lenin, and the official program of the Communist Party of the Soviet Union, Zvorykin seeks to reconcile the orthodox determinism of historical materialism (the view that society—and technology—develops solely in response to economic needs and according to objectively existing laws, which operate independently of human will) with the fact that technology is, after all, developed by people according to various subjective aims they set themselves. But the economic and social relations that have been created in the USSR are said to guide the Soviet technologist in the right direction; unlike capitalist society, which turns technology against mankind, socialism—and only socialism—uses technology for the well-being of the people.

Finding a "struggle of opposites" in machine technology is not a simple task even for a Soviet scholar. The best he can do is to point to what engineers call "trade-offs" between various operating characteristics (for example, power vs. utilization factor—the more powerful a machine, the less likely it is to be used to full capacity) and to the law of diminishing returns (productivity does not increase indefinitely with the increase in any one factor of production but tends to level off). It is obviously stretching things a bit to call such trade-offs "internal contradictions."

Zvorykin next turns to the highly theoretical question of whether science is itself a productive force or merely one of the forms of social consciousness that contributes to the development of productive forces. He argues for the first view, by citing the great modern development of the several sciences, their growing interdependence, and the decisive role played by science in the expansion of technology. Why does he pay so much attention to what is, after all, a rather academic question? The reason is that Lenin considered industrial production a criterion of social progress. Counting science as an element of production thus makes it possible to capitalize (if that is the word) on the undoubted successes of Soviet scientists. Such an approach may also affect the organization of Soviet scientific research and its interaction with industry. The Second Industrial Revolution has not bypassed the USSR, and its scientists and engineers are just as fascinated with the possibilities of atomic energy, materials processing, electronics, automation, computers, and cybernetics as their Western colleagues—matters that Soviet politicians doubtless view just as suspiciously as *their* Western counterparts.[29]

What of the change in man's place in the production process? As first production itself and later even its control are automated, says Zvorykin, man will be freed from onerous labor and will have more leisure, which will give him, in Marx's words, "the time for loftier activity." The problem of the value of labor in a fully automated world can be only solved under socialism, since "the conditions in an exploiter society are such that the reduction of the working hours entails reduction of wages." (He gives no details about the socialist solution, beyond saying that under socialism the new leisure will be absorbed in social and scientific activity and cultural and physical development.)

The orthodox view of technology is thus highly optimistic. Socialism is fully capable of protecting the people against the technocrats and of using technology for humanitarian ends, and as everyone is brought up according to a highly idealistic program of "social knowledge," no degradation of human values (such as proclaimed by Ellul) is possible. All is for the best in the best of all (communist) worlds.

We have dealt with the orthodox Soviet view at some length because the USSR has the largest industrial production among the communist countries and the most monolithic ideology, which it rigorously imposes on its dependents. China is still developing as an industrial power, and if the indications along the way are portents for the future, it is likely to develop a somewhat different viewpoint. The Chinese leaders are making conscious efforts to avoid such known disadvantages of industrialized societies as urban sprawl and the excesses of large-scale automotive transport. If any workers are released from the land they are not allowed to migrate toward a few metropolitan areas but are kept in provincial towns, and virtually no one has a car at his disposal. (Keeping down concentrations of a dispossessed agrarian proletariat and the mobility of the population also happens to have clear political advantages.) Of the countries within the Soviet orbit (not counting Yugoslavia, an imperfectly controlled society from the Soviet viewpoint), all have pretty well fallen in step. It is instructive to note what happened in one that attempted to go its own way.

In 1967, a first-rate team of Czechoslovak sociologists, economists, philosophers, and other specialists published a study of the effects of the Second Industrial Revolution entitled *Civilization at the*

Crossroads.[30] The study was a serious attempt to reconcile the obvious effects of a development Marx could scarcely have foreseen with his writings. That was not an easy job. Marx had held that industrialization rationalizes work to break it up into simple, monotonous tasks; that the growth of society's productive force can be best served by concentrating all resources in industry; and that man's main contribution to that growth lies in his value as a unit of labor. It now appeared that exactly the opposite is true: that automation does away with unskilled labor, that the vital measure of production is the amount of resources released to the cultural and social sectors, and that harnessing man's creative powers (notably in science and technology) is vastly more important than using him as a worker. There was an unseemly scramble to find the right quotations in Marx's works to support the new findings. Especially painful for the authors was the grudging recognition that modern technology demands a reorganization of society in both capitalist and communist countries, and that on the whole the capitalists seem to be doing a rather better job of adaptation.

As an example, it was generally conceded to be a difficult task to motivate socialist workers to exert themselves as hard for the common good as capitalist workers do for profit. Would breaking down national enterprises into local ones help? Could one introduce competition among enterprises, let supply and demand be shaped at least in part by the market rather than by central planning, and allow the workers partial participation in the profits of their collectives? But such ideas sounded dangerously like capitalism. The authors were quick to protest that they were not *really* falling back on the profit motive, that under socialism all individual interests are submerged in the general interests of each and every enterprise and indeed of the entire community. What's good for socialism is by definition good for the worker. But it was not good enough for Moscow, where these tentative gropings, the first breath of fresh air in Marxist thought in half a century, were perceived as clear warnings of heavy weather ahead. The Czech comrades would be wanting to reform the party next! In fact, such a thing had actually occurred to them, but conscious of Big Brother next door, they had managed by and large to keep overt political debate out of it; or so they thought. But ideas on technology cannot be divorced from politics, as we have been at some pains to show.

Civilization at the Crossroads is thus interesting for at least two reasons. First, it is the communist intelligentsia's most advanced study of the effects of technology, going far beyond the relatively tame Soviet strictures. Look, it seems to be saying, this thing is bigger than capitalism or communism. Modern technology provides opportunities for the development of man undreamt of in your philosophy. We must make the most of it, and if we succeed, we shall carry our society and the human potential far beyond anything the world has seen. Beside such a vision, the differences between present-day social systems pale into insignificance.

The second reason for the work's importance is that it played a leading part in the momentary revival of progressive thinking about social matters that accompanied the "Prague Spring" of 1968, that outflowing of hope that ended so tragically in Soviet armed intervention. In their concern that more modern attitudes toward technology and the management of Czechoslovak industry would have repercussions on the organization of the communist leadership and armed forces of that country, the Soviet leaders could think of only one thing to do with the Czech revival: nip it in the bud by armed force. In 1968, they ordered an invasion, occupied the country, and forced massive changes in the Czechoslovak government and intellectual establishment. Thoughtful observers estimated that if the Czechs had been allowed to put their reforms across, the result would have been a strengthening of the communist cause rather than otherwise; but to make the reforms general, the Soviet leadership would have had to step aside in favor of a more liberal regime, and that is something it is unlikely to do voluntarily. The episode thus remains an object lesson that philosophers can indeed affect the course of history (the editor, Radovan Richta, was director of the Philosophical Institute of the Czechoslovak Academy of Sciences)—but that they are unlikely to prevail in the face of a well-organized establishment backed up by armed force. The Prague Spring was of short duration, a mere crack in what continues to be the monolith of official attitude toward technology in the Soviet Union and its satellites.

Double electrified fence surrounding Auschwitz extermination camp.

Chapter 7 Technology as a Social Force and Ethical Problem

TECHNOLOGIST: BENEFACTOR OR MONSTER? is the question we may well ask after thinking about the several views of technology described in the previous chapter. Along with the many benefits that technology has showered on those who have had the good luck to profit from them, it has also brought monstrous dangers: the possibility of nuclear warfare, overpopulation, the dehumanization of some people by mass society, the despoliation of the natural environment we claim to control. Should we allow technology's practitioner, the technologist, to force his will on society, or would we be better advised to stay his hand? Or is that not the problem: is the technologist an insignificant agent acting at the behest of societal forces far mightier than he?

The truth doubtless lies somewhere in between. The technologist is neither the master of our fate nor the helpless pawn of an inexorable historical process. In fact his role varies from case to case. The engineer who develops a new system of color television can scarcely be blamed if tasteless programs are shown; he has no control over programming. But the responsible engineer who plans a river dam to provide a source of hydroelectric power would consider his design incomplete if it did not also detail an estimate of the cost of the produced electricity in comparison with that from alternate sources, the effects of the project on irrigation and flood control, the recreation opportunities in the lake created behind the dam, and its topographical and ecological effects.

The recognition that an engineer's responsibilities extend beyond technical and economic considerations is long established and antedates more recent popular concerns with environmental and similar problems. Writing in the 1929 edition of the *Encyclopaedia Britannica* (on "Engineer, Professional"), Alfred Douglas Flinn said:

> The engineer is under obligation to consider the sociological, economical and spiritual effects of engineering and operations and to aid his fellowmen to adjust wisely their modes of living, their industrial, commercial, and governmental procedures, and their educational processes so as to enjoy the greatest possible benefit from the progress achieved through our accumulating knowledge of the universe and ourselves as applied by engineering. The engineer's principal work is to discover and conserve natural resources of materials and forces, including the human, and to create means for utilizing these resources with minimal cost and waste and with maximum useful results.

Just how far does the engineer's responsibility extend? In the example mentioned previously, the planning of a river dam, should he be also held responsible for inequities that may arise in the resettlement of inhabitants from the area to be flooded? Should he concern himself with the scheduling of the distribution of water for irrigation purposes? Should he make it his business to see that a part of the new lakeshore is set aside for public parks? Or are these political questions, beyond his competence and not really amenable to technical solutions? When a public structure—a highway, a government building, or a state university—is planned, the state may invoke the legal principle of "eminent domain" to take over private property for reasonable compensation (or else a single, stubborn property owner could hold up the state, in more senses than one); but a disgruntled property owner may still blame the engineer who planned the structure for any personal inconvenience and injustice that results.

The question of the technologist's responsibility for the consequences of his works thus appears to be enormously complicated. The unforgiving critic of technology puts full responsibility for all ramifications of technical inventions on their originators. The apologist sees technology as a means that society is free to use or not, as it chooses; according to this view, technology opens doors but does not compel men to enter. Yet that view is clearly disingenuous. It is expecting too much of man not to enter a door that has been invitingly opened for him, if only to find out what lies beyond. Moreover, society might well exercise some control over which doors to open, of assessing the effects of technological decisions beforehand—a problem that we take up towards the end of the next chapter. But perhaps the engineer should voluntarily restrain himself and act only in accordance with a strictly conceived code of ethics, one that would go beyond the well-established code governing the professional conduct of engineers to something approaching the principles to which physicians subscribe? Before we come to such an Engineer's Hippocratic Oath, we shall try to show that the technologist can be given neither full credit nor full blame for developments which at first blush appear to be largely technical in nature; society at large must answer for both good and bad. We shall illustrate the point by two cautionary tales, cases taken from recent history, one benign in its effects and one malignant.

AGRICULTURAL EXTENSION, *the network of central services and county agents that provide leadership and carry information to farming communities throughout the USA,* is widely given credit for the high efficiency of American agriculture. Started in the early 1900s in a predominantly agricultural land, the services of the county agents (whose work received federal recognition and support with the passage of the Smith-Lever Extension Act in 1914) quickly found a place in rural America.[1] They worked in cooperation with the agricultural colleges that had sprung up everywhere as a result of the passage of the Morrill Land-Grant College Act in 1862 (chapter 2), as well as the agricultural experiment stations set up in most states after the Hatch Act (1887) had provided federal subsidies for agricultural research. Today the Cooperative Extension Service is financed jointly by the U.S. Department of Agriculture, by the states, and by local agencies, and has a professional staff of over 15,000, with local volunteers numbering more than 1 million in over 3000 counties. It all started with one community demonstration farm in Terrell, Texas, in 1903.

The man who started the system was a country teacher and farmer who acquired a technological education largely through his own efforts. Seaman Asahel Knapp (1833–1911), born of pioneer colonial stock on an upstate New York farm, was well educated by the standards of the day.[2] He attended Troy Academy in Vermont and Union

College in Schenectady, N.Y., whose president was the remarkable pragmatist Eliphalet Nott (1773–1866), an inventor and educational innovator, one of the unsung heroes who pushed for a more widespread and more practical higher education in a day when the only college curriculum was the classical. After teaching for 10 years, Knapp injured his knee and on medical advice went back to the open-air life on a farm. He moved west, to Iowa, but suffered several setbacks because farm conditions there were different and no information for new settlers was available. He laboriously educated himself in scientific farming methods, and when he regained use of his leg in 1875, he determined to try out his newly acquired knowledge. He became an exceptionally successful breeder of pigs. By paying close attention to circulars that had begun to come from the new U.S. Department of Agriculture, he created a model farm that supplied brood stock to other farms all over the Middle West.[3] At the same time he became devoted to the idea of disseminating scientific farming methods with an almost religious missionary zeal. He helped to start the *Farmer's Journal* in 1872, became a frequent contributor, and finally the editor. He advocated setting up agricultural experiment stations and took part in the lobbying that led to the Hatch Act.[4] He spoke tirelessly to granges, breeders' and farmers' associations, and other groups. But he really hit his stride when he was appointed, in 1880, Professor of Practical and Experimental Agriculture and superintendent of the college farm at the Iowa State Agricultural College in Ames.

That institution (now the Iowa State University of Science and Technology) had been founded in 1869 as one of the land grant colleges. Under Knapp's leadership (he also served as president from 1884 to 1886) it became a center for teaching "a science in agriculture as distinct from the sciences related to agriculture." He created a new type of curriculum, leading to the degree of Bachelor of Scientific Agriculture. He tried to tell everyone about the importance of research, but there he was less successful. At a time of rising land prices farmers could turn a profit more quickly by real-estate speculation than by improving their farms. The total annual Iowa appropriation for research in agriculture and horticulture was $1500, and the college administration was subject to every political wind or even zephyr that blew from the state capital. Finally, Knapp had enough. In 1886, at the age of 53, he resigned and turned to an entirely new venture: supervising a large reclamation project in Louisiana to clear tidelands for the cultivation of rice, jute, and vegetables. For the next 17 years, he was involved in one agricultural venture or another in the South, traveling tirelessly all over the world, towards the end as a government plant explorer to study rice varieties in the Orient.

Working with Beverly Thomas Galloway (1863–1938) of the U.S. Department of Agriculture, Knapp next attempted to set up demonstration farms in several Gulf states in which the economic advantages of crop diversification and scientific farming would be made obvious. He soon found that showing how to do it on a government model farm was not enough, since a small farmer could not identify with an enterprise that was backed by the seemingly infinite resources of the federal government. The solution was the community demonstration farm, a scheme first tried in 1903 on Walter C. Porter's farm in Terrell, Texas. Eight Terrell businessmen and farmers were persuaded by the now 70-year-old Knapp to subscribe to a $450 indemnity fund to protect Porter against possible loss resulting from experimentation with new crops, fertilizers, and planting methods on 70 acres of his 800-acre farm; but he was to keep all profits if he came out ahead. In fact he came out $700 ahead on the 70 acres and announced that he would work his entire farm according to the new principles the following year.

The idea soon spread. Farmers throughout the area clamored for similar treatment. Almost inadvertently, Knapp had stumbled on the proper formula: keep the government on the sidelines, as an advisor, and motivate the farmer to make good in the eyes of his friends and neighbors by letting him reap full credit for any improvements resulting from innovation. Then a new challenge arose: an infestation of the boll weevil, an insect that threatened to wipe out cotton farming throughout Texas in 1903–1904. Knapp's demonstration method was promptly adapted to teaching the various remedies that had been worked out by government entomologists to fight the pest and his agents were able to stop the infestation wherever they went.

The agents thus emerged as the kingpins of the entire demonstration system. Financed at first entirely by local tax support and private contributions (notably $1 million from the General Education Board funded by John Davison Rockefeller, 1839–1937), and introduced in northern and western states by Galloway and by William Jasper Spillman (1863–1931), the county agents qualified for federal sponsorship with the passage of the Smith-Lever Act in 1914. By 1917, when the USA entered World War I, there were 1466 agents; their number was soon nearly tripled, partly by appointments of emergency agents made in the nationwide effort to feed America and her allies.[5] After the war, the county agent broadened his horizon beyond farm efficiency to such innovations as better drainage and irrigation, the control of plant and animal diseases, and advice giving on how to remodel homesteads and form marketing associations. He helped organize farm boys in 4–H clubs and girls in canning clubs.[6] With the flood of new government programs that followed the depression of the 1930s—control of soil erosion, deliberate reduction in the output of some farm products, resettlement, electrification, crop insurance—the county agent also became an interpreter of the new legislation and a long-range planner. He met yet another challenge during World War II, when agricultural production in the USA again rose sharply despite shortages in manpower and in farm machinery. Since then his concerns have again shifted, with increasing mechanization, to new managerial and marketing methods in what has come to be called "agribusiness," and to social, economic, and political problems arising from the inability of some small farms to hold their own in the face of the growing rationalization of agriculture.

The county agent thus emerges as the technologist par excellence, and the man who thought of the scheme as a great benefactor of mankind. Experts from all over the world come to study the system with a view to introducing it in their countries.[7] The unmatched productivity of American farms is ascribed to the county agents' efforts. There can be no nobler calling than one that does so much toward fulfilling the greatest need of one's fellow beings and that holds the promise of banishing hunger from the face of the earth.

EUTHANASIA *is the practice of painlessly putting to death persons suffering from an incurable condition or disease.* Anyone who has seen an aged relative or friend through a painful, terminal illness has wanted to stop the agony and to grant the sufferer the release of a painless death at the hands of a physician acting with the approval of the patient's family. Yet most thoughtful individuals, even when not held back by religious scruples (such as the prohibition against killing contained in the biblical Sixth Commandment), are appalled at the ease with which the practice of euthanasia can be perverted for criminal ends, as was amply shown by the Nazis during their short but diabolic rule—the twelve years from 1933 to 1945, during the second half of which they waged a world war whose outstanding feature was the huge number of civilians killed.[8]

Early in the war the Nazi leader, Adolf Hitler, asked his personal physician, Karl Brandt, to start a euthanasia program directed against deformed children, the chronically ill, the incurably insane, and the aged. These "useless eaters" not only took up hospital space and medical services (including doctors and nurses) that would soon be at a premium, but they were also a source of embarrassment to the Nazi theory of the perfect master race. Patients in the condemned categories were selected from nursing homes, hospitals, and asylums by government teams under the direction of Dr. Herbert Linden and were moved to collecting centers and finally to euthanasia stations, where they were executed, usually by intravenous injections of carbolic acid, though shooting with a revolver was also recommended by the police officer in charge, Christian Wirth.

Such killings were not efficient and they could not be kept secret. The relatives, who had of course not been consulted, became suspicious of the stereotyped wording of the falsified death notices, sent from strange locations. The arrival of large batches of patients and the many cremations could also not be wholly camouflaged. Over 275,000 aged, insane, and incurably ill had been done away with before the program was shifted away from civilian facilities in late 1941, mainly because several churchmen had protested openly (some of them from their pulpits) that the notoriety of the practice had a demoralizing effect on the population and undermined the concept of authority.

The fact that the energetic intervention of religious leaders in a domestic euthanasia action carried out against their coreligionists brought the action to an end shows that not even a highly oppressive, totalitarian government with wartime powers and a subservient press can afford to ignore the pressure of suitably channeled public opinion. The pity is that virtually no such opposition found expression when the Nazis placed the action under the control of party formations, put it out of sight in concentration camps, and finally shifted it to occupied territories and directed it against populations of other religions and nationalities in what was to become the greatest organized massacre in history.[9]

The party organization put in charge of the euthanasia program was the paramilitary *Schutz-Staffel* (defense echelon), or SS for short, and in particular the elite SS Death-Head units (whose members wore the dread skull-and-crossbones insignia), which ran the concentration camps and provided the technicians who were to engineer the mass exterminations. Henceforth, euthanasia victims would not be selected from among incurables but exclusively from concentration-camp inmates *unfit for work*. Inmates were selected for "invalid transports" by a commission that never saw the prisoners. "Merely paper work," wrote Dr. Mennecke, one of the commission members, to his wife from the concentration camp at Buchenwald near Weimar. (Both Mennecke and Brandt were hanged after the war for their part in the euthanasia program.) In 1943, a further directive restricted the euthanasia program to insane persons only. "All other prisoners unfit for work are to be absolutely excluded from this operation," the new order stated. "Bedridden prisoners are to be given suitable work, such as can be done in bed." But by this time, the "invalid transports" had become relatively unimportant, for the vastly larger program of exterminating members of inferior races was going at full blast.

With the occupation of the Soviet-held half of Poland after Hitler's surprise attack on the USSR in 1941, the Nazis got a chance to put into effect their long-range plan of exterminating all people considered racially and biologically inferior (the so-called *Endlösung*, or Final Solution) and removing all incorrigible political opposition. Since millions of persons were involved, the technical problems of secretly assembling and killing them and getting rid of the bodies were staggering. The first exterminations—of the non-Aryan population of the smaller Polish towns and of Communist party members and political commissars among Soviet prisoners of war—were relatively simple: the victims were marched out of town, told to dig a long ditch, and then shot in the back of the head at the edge of the ditch so that they fell in, layer upon layer, until the mass grave was full. (The Soviet poet Evgeny Yevtushenko's famed work, "Babi Yar," refers to the particularly large execution carried out in this way in a ravine of that name near Kharkov in the Ukraine.) But the system was clearly unsuited to mass exterminations, for reasons succinctly summarized by the French writer Jean-François Steiner:

> The method of shooting in itself gave rise to controversy among the [SS] technicians, who were divided into two schools: the "classics" and the "moderns." The first were advocates of the regulation firing squad at twelve paces and the *coup de grace* given by the squad leader. The second, who felt that this classic apparatus did not square with the facts of the new situation, preferred the simple bullet in the back of the neck. The latter method finally prevailed, because of its efficiency. It was here that the psychological problems vividly emerged.
>
> With a firing squad you never knew who killed whom. Here, each executioner had "his" victims. It was no longer squad number such-and-such that acted, but rifleman so-and-so. Moreover, this personalization of the

act was accompanied by a physical proximity, since the executioner stood less than a yard away from his victim. Of course, he did not see him from the front, but it was discovered that necks, like faces, also individualize people. This accumulation of necks—suppliant, proud, fearful, broad, frail, hairy, or tanned—rapidly became intolerable to the executioners, who could not help feeling a certain sense of guilt. Like blind faces, these necks came to haunt their dreams. Paradoxically, it was from the executioners and not from the victims that the difficulties arose. Hence, the technicians took them seriously.

Thus there arose, no doubt for the first time in the world, the problem of how to liquidate people by the millions. Today the solution seems obvious, and no one asks himself the question. In 1941, it was quite otherwise. The few historical precedents were of no use, whether it was a question of the extermination of the Indians by the Spaniards in South America or by the Americans in the United States, or again of the Armenians by the Turks at the beginning of this century. In these three cases, no attempt had been made at a new technique, no advance beyond the time-honored hanging and shooting, which, as we have seen, did not satistfy the technicians.

It was necessary to invent a killing machine. With a methodical spirit that is now well known to us, the technicians defined its specifications. It had to be inconspicuous to avoid arousing anxiety in the victims or curiosity in the witnesses, and efficient enough to be on a par with the great plans of the originators of the Final Solution; it had to reduce handling to a minimum; and finally, it had to assure a peaceful death for the victims.[10]

The killing method that ultimately came to be used was one that had been tried in a tentative way during the euthanasia program: gassing by carbon monoxide. Engine exhaust was piped into the back of a hermetically sealed van loaded with 15–20 victims. The technique was clearly inadequate when it came to really large numbers, such as the 400,000 inhabitants of the Warsaw ghetto (whose fate has been so graphically described in John Hersey's 1950 novel, *The Wall*). That problem was solved by backing the van against a sealed building and piping the exhaust gases into it. Brought to assembly-line perfection by a young SS lieutenant named Kurt Franz at the Treblinka extermination camp near Warsaw, the system ultimately handled 2600 victims in 13 gas chambers with a 200-person capacity in 45 minutes, including the time needed to strip them, cut off the women's hair for use in the war economy, and removing wedding rings and extracting gold teeth from the corpses.

That still left the problem of disposing of the bodies. At Treblinka, the first 700,000 were buried in the usual ditches before the chief of the SS, Heinrich Himmler, decided after an inspection that they would have to be disinterred and burned—a gigantic task that no one had anticipated at the start of the euthanasia program. Even with earth-moving machinery and an unlimited labor supply, it was hard to see how more than 1000 bodies could be handled per day, a rate at which it would have taken 700 days or nearly 2 years to do the job—even if the putrid corpses had not proved so very difficult to ignite. At that point an SS specialist in cremation was summoned, Herbert Floss, who had perfected a new technique at various smaller concentration camps and was itching to try it out on the really large scale that Treblinka afforded. He constructed a giant grill of railroad rails placed on concrete supports about one meter off the ground, on which the bodies were piled in layers and burnt in open air. The giant funeral pyres were an immediate success and the grisly job of excavation and burning began at once. It did not stop for several months.

In the meantime, the largest extermination camp of all had arisen in southern Poland at Oświęcim, known in most non-Slav languages as Auschwitz.[11] The commandant was an SS lieutenant colonel named Rudolf Hoess (not to be confounded with Hitler's deputy Rudolf Hess, who had flown to enemy Britain in 1941 in an alleged attempt to negotiate a peace treaty). Hoess made substantial improvements in killing technique (notably the use of prussic acid in place of carbon monoxide in the gas chambers) and in the methods of disposing of the corpses. Even Herbert Floss's efficient funeral pyres were inadequate for a plant that was to produce more than 4,000,000 bodies; crematories were needed. Also, for real efficiency, the reception hall (disguised as the dressing room of a public bath), gas chamber, and crematory were combined into a single building.

Further refinements were introduced later, including peepholes, tracks and electric elevators to convey the bodies, and special metallurgical furnaces in which dental gold was melted down into ingots of standard shape and size. Four plants were ultimately built at Birkenau near Auschwitz, two large ones and two slightly smaller. Each was surmounted by a large chimney. Grouped around the chimney of each of the two large crematories were nine furnaces, not unlike the blast furnaces used in steel mills (chapter 2), fuelled by coal and fanned by electric blowers. Each furnace had four openings. Three corpses could be placed simultaneously in each opening. Thus, 108 bodies could be burned by one crematory in a single operation, and about 360 by all four crematories. It took about half an hour to reduce a body to ashes—720 per hour, or 17,280 each 24 hours, for the furnaces were often in operation day and night. In an emergency, several of the old open-air pyres could be also pressed into service. Peak production occurred on 29 June 1944, when over 24,000 people were gassed and burned in one 24-hour period.

DIVISION OF RESPONSIBILITY *between the technologist and the society that uses him cannot be reduced to a formula.* In the two examples that we have described above in some detail, the assignment of credit and blame might appear to be a very simple matter indeed. Seaman A. Knapp, technologist extraordinary, introduced the county agent system of agricultural extension by which the application of agricultural research to American farm production caused it to rise to unmatched levels, so that the USA became the world's breadbasket and saw its methods copied by newly developing lands all over the globe; he must receive full credit for this splendid achievement. The technicians of the SS, Wirth, Franz, Floss, Hoess, and the rest, hold full responsibility for the horrors perpetrated by the Nazis, which they alone knew about and could have stopped.

Yet the objective historian would not see the matter in such simple terms. He would point out that Knapp was not really a technologist; except for three short college courses—on electricity, mechanics, and astronomy—his formal preparation was entirely in the classics and his brief tenure at a newly founded prairie college was based mainly on his reputation as a successful stock breeder and editor of a farm journal. Moreover, other factors, independent of the county agent system, probably had more to do with the success of American agriculture, such as unrestricted immigration, the Homestead Act of 1862 (by which the title to 160 acres of unoccupied land is transferred to any person who undertakes to improve it on payment of a nominal fee after five years of residence), the spread of the railroads, the rationalization of production and of distribution and a host of other developments quite unconnected with education and the dissemination of information. Viewed in this light, the unique contribution of the county agent system lies rather in the social and political realms, through enabling small farms to survive (which is not necessarily the best outcome from a purely technical viewpoint) and their owners to organize into viable production and marketing associations.

The responsibility for not stopping the Nazi holocaust might also be placed by the objective observer elsewhere than on a small group of technicians. Quite apart from the mass responsibility for allowing a group of criminals to take over a country's government, one cannot shuttle millions of people all over a continent and create a complex undertaking for disposing of them and of their possessions without sharing knowledge of it with tens of thousands of people. Beyond the SS and the army, there were the salaried workers, foremen, and managers of the industrial plants; the contractors who supplied materials and elevators, blowers, and furnaces for crematories; the railroad men who assembled and ran the deportation trains; and the welfare agencies that handled the distribution of the shoes and clothing from the dead. An SS general, Oswald Pohl, head of the SS supply and administration headquarters that was responsible for all concentration and extermination camps, admitted at Nuremberg that it was not true that only a handful of men in his organization knew; "in the case of textiles and valuables," he said, "everyone down to the lowest clerk knew what went on in concentration camps."[12]

Furthermore, none of the men named was really a technologist by training or occupation. Wirth, the manager of the first euthanasia program, was a policeman, superintendent of Stuttgart's criminal investigation division. Franz, in charge of prisoners at Treblinka, had been a waiter in a small town in Bavaria. Floss, the specialist in open-air cremation, was entirely self-educated. Hoess, the commandant of Auschwitz, had been a professional SS officer since 1934. In fact, except at the very beginning of the Nazi ascendancy, the SS had had difficulties all along in trying to recruit professional men. When the SS got hold of an engineer, they hung on to him. For instance, Kurt Gerstein, a mining engineer, had joined the Nazi party in 1933 but had subsequently been imprisoned and expelled for part-time religious activities that were held to be inimical to the new regime. When he asked for reinstatement after the war broke out, he was not only readmitted but given an opportunity to join the SS and assigned to its sanitation service, over the protests of party regulars. Moreover, there are strong indications that he deliberately joined because he wanted to find out more about the rumored euthanasia program—a resolve that was strengthened when his own chronically ill sister-in-law fell victim to it and which led him to embark on the extremely dangerous journey of active opposition.

Among students of the few exceptions to the monolithic blind obedience characteristic of the Nazi era, the case of this courageous engineer—the only SS officer known to have attempted to interrupt the extermination program by making its existence known to the world—has attracted much attention. A somewhat idealized portrayal of him appears in Rolf Hochhuth's play *The Deputy.* Archives of materials pertaining to Gerstein (who committed suicide in a Paris prison in July 1945, before his case had been fully sorted out from among those of the many war criminals then held by Allied authorities) have been established in the Kurt-Gerstein-Haus in Berchum near Hagen in Westphalia. Several book-length studies about him have been published, the most penetrating of which is Saul Friedländer's *Kurt Gerstein: The Ambiguity of Good,* which explores the nightmare world of a practicing Christian devoted to the self-imposed mission of arousing his countrymen and the world to the horrors of the extermination program.[13] Because of his expertise with minerals and chemicals, Gerstein, his worst suspicions about the euthanasia program quickly confirmed, soon found himself at its very heart: he was charged with disinfecting the clothing collected at the extermination camps and supplying the gas chambers with prussic acid. Devastated by his first visit to the extermination camps, Gerstein lost no time in blurting out the whole story to a Swedish diplomat he happened to meet on the train back to the capital, who duly passed it on to his neutral government, which in turn communicated the information to the British Foreign Office. Gerstein also made extensive disclosures to a fellow engineer serving in the air force, Armin Peters, and later to a Dutch colleague, the engineer H. J. Ubbink; he attempted to inform the Vatican legation, which threw him out; and spoke to many others, mainly Catholic and Protestant leaders and neutral diplomats. Gradually the story seeped abroad, confirming other accounts that were reaching the Allies and the Vatican from various sources. But no disclosures were made until the last year of the war, and no domestic protest could be organized.[14] Gerstein's effort remained an "appeal without echo," in Friedländer's words, "and Gerstein himself, an impotent witness, was caught up in the wheels of the machine he was trying to halt."

Our two cautionary tales, taken from opposite ends of the moral spectrum, thus show the futility of trying to apportion credit or blame for the effects of technological operations between society and those it employs to carry them out. But has not the technical expert, because of his special knowledge, a duty to do more than to make an objective presentation of the various solutions to a problem and to draw up balance sheets of the technical advantages and disadvantages of each? Should he sometimes set aside his scientific objectivity and make a special plea for one solution or another on the basis of nontechnical considerations? Will society be better or worse served if its decision makers cannot rely on the objectivity of the technical information on which they must base their decisions?[15]

What is evidently wanted is a set of balance sheets in which the relative merits of each solution to a technical problem are analyzed both on technical grounds such as safety, ease of operation, cost, reliability, maintainability, complexity, and esthetics; and on ethical grounds such as moral considerations, effects on the quality of human life, liberty, dignity, and other human values. How to weigh each of these attributes against the others remains an unsolved problem. The engineer is quickly out of his depth[16] and can justly claim that even minds generally assumed to be better suited than his to making moral judgments—philosophers, religious leaders, political and other social scientists—do not universally agree on ethical criteria for human action. One envies the framers of the American Declaration of Independence their easy assurance to "hold these truths to be self-evident, that all men are created equal, that they are endowed by their Creator with certain unalienable Rights, that among these are Life, Liberty, and the pursuit of Happiness." It was an excellent program and one that has stood the American republic in good stead, but that did not stop them (any more than it did the ancient Greeks, whose civilization is still held up as the model of a society built on human values) from enslaving their fellow men, a fact that we find abhorrent and immoral. Not even established religion is impervious to change. The last century alone has seen tremendous changes in the outlook of the Roman Catholic Church, for instance, with such papal encyclicals as *Rerum Novarum* (1891) of Leo XIII, *Quadrigesimo Anno* (1931) of Pius XI, and *Mater et Magister* (1961) of John XXIII; and Protestant theology has not exactly stood still, either, in the hands of a Reinhold Niebuhr (1892–1971) or a Paul Johannes Tillich (1886–1965).

Values change, if only because they depend on knowledge; and conversely the advance of knowledge depends on values, for example the value of free dissemination of knowledge and the value of truth. Yet some critics minimize this interdependence and hold that to base value judgments on knowledge—even knowledge of history, say, or anthropology—is dangerous if it leaves out intuition and tradition (to say nothing of divine revelation). Still others fall back on the moral law embodied in the categorical imperative of Immanuel Kant (1724–1804): Act only as if the maxim from which you proceed were to become, through your will, a universal law. That seems a perfect prescription for the absolutely good will of a rational being, and moreover one independent of theological considerations; yet even well-meaning men do not always act as they ought, but rather according to inclination.

Pity then the poor engineer, who must pick his way through a thicket of contradictory and changing philosophical systems, holding now intuition, now experience or conscience supreme, and either the state or religion or some other external authority as the ultimate arbiter of individual conduct.

Yet a set of criteria based on common sense cannot be all that difficult to construct. Consider a tentative program, a list of values that a distinguished American jurist and civil servant, A. A. Berle, Jr., got off in no more time than it took to write a magazine article:

1) People are better alive than dead.

2) People are better healthy than sick.

3) People are better off literate than illiterate.

4) People are better off adequately than inadequately housed.

5) People are better off in beautiful than in ugly cities and towns.

6) People are better off if they have opportunity for enjoyment—music, literature, drama, and the arts.

7) Education above the elementary level should be as nearly universal as possible through secondary schools, and higher education as widely diffused as practicable.

8) Development of science and the arts should continue or possibly be expanded.

9) Minimum resources for living should be available to all.

10) Leisure and access to green country should be a human experience available to everyone.[17]

Few would find fault with these propositions. They do not constitute a specific program or deal with such ethical problems as prejudice or dishonesty, but anyone looking for some general guidelines by which to assess the advisability of a new technological undertaking could do worse for a start than to consider it against nonquantitative criteria such as the above. They are certainly more useful in that regard than the American engineer's Code of Ethics, which was got up by the several engineering societies mainly to govern the relations of "professional" engineers (that is, individuals or partnerships that offer professional services in much the same way as lawyers) with each other and with their clients. For the engineer himself, something on a loftier moral plane is needed.

AN ENGINEER'S HIPPOCRATIC OATH *can be developed in analogy with the principles attributed to the school of Hippocrates of Cos* (c. 460–c. 370 B.C.), the great Greek physician who first separated the practice of medicine from superstition and philosophy and based it on observation and reason. Like medicine, engineering is above all concerned with improving the human condition, and there are other parallels between the two professions as well, both in the preparation of their practitioners and in their practice. An oath based on Hippocratic teachings is administered to graduates of many modern schools of medicine. An amended version of it is suggested here for engineering graduates:

AN ENGINEER'S HIPPOCRATIC OATH

I solemnly pledge myself to consecrate my life to the service of humanity. I will give to my teachers the respect and gratitude which is their due; I will be loyal to the profession of engineering and just and generous to its members; I will lead my life and practice my profession in uprightness and honor; whatever project I shall undertake, it shall be for the good of mankind to the utmost of my power; I will keep far aloof from wrong, from corruption, and from tempting others to vicious practice; I will exercise my profession solely for the benefit of humanity and perform no act for a criminal purpose, even if solicited, far less suggest it; I will speak out against evil and unjust practice wheresoever I encounter it; I will not permit considerations of religion, nationality, race, party politics, or social standing to intervene between my duty and my work; even under threat, I will not use my professional knowledge contrary to the laws of humanity; I will endeavor to avoid waste and the consumption of nonrenewable resources. I make these promises solemnly, freely, and upon my honor.[18]

We have seen that much of the responsibility for the uses of technology lies with society. Blaming engineers for the shortcomings of technological society makes about as much sense as blaming the failure of a new play on the stagehands. On the contrary, we may come to look to the engineer for moral guidance. If the engineer subscribes to principles such as the above, the leaders and representatives of the rest of society can do no less. Only by the concerted efforts of all can the abuses of technology be avoided. We may paraphrase Clemenceau and say, technology is too important to be left to the engineers.[19]

A polar map of the world surrounded by olive branches is the emblem of the United Nations, whose technical agencies have led the way toward international cooperation.

Chapter 8 Challenges

INTERNATIONAL UNDERSTANDING, *the road to peace, is the greatest problem facing mankind.* Technology can contribute much to its solution. Like science, technology knows no national boundaries; and technology is foremost among the gifts that the rich nations can bestow on the poor to begin to close the gap and lessen world tensions.

The international nature of technology is exemplified by the many technical agencies of the United Nations organizations that have continued to function without hindrance through crises, confrontations, "cold" wars, and wars not so cold. Old hands at the U.N. and its predecessor, the League of Nations, are given to saying that if only nations could get along in all areas as well as they do in technical matters, the world's greatest problem—the search for peace—would soon be solved. Specialized agencies concerned with health, aviation, weather telecommunications, postal services, and similar matters have been running at a level of international cooperation that turns their counterparts on the social, economic, and political side green with envy. Some of these agencies are much older than the United Nations organization itself. The Universal Postal Union was founded at Berne in 1875. A predecessor of the

International Telecommunications Union at Geneva, the International Telegraph Union, was founded in 1865; it is the oldest existing international administrative organization and its roots go back even further, to the Prussian-Austrian treaty of 1849 and the West-European Telegraph Union of 1855. Until about 1851, a telegram from a French or Belgian town to a German one would be wired to a border telegraph office, transcribed, and carried to the office across the border for forwarding, sometimes after having been first translated. Such absurdities are intolerable from a technical viewpoint and international agreements were not long in coming. To say that the ease with which purely technical agreements can be made derives from the fact that their framers are often themselves technical men or draw their inspiration from them may seem extravagant. Nevertheless, here is what the chairman of the first post–World War II Radio Conference said to the delegates: "We are challenged to match our skills as conferees with the superb skill of the scientists who have given us the many wonders of radio. To achieve that goal we must rise to a high order of statesmanship."[1]

What technology can do for developing countries has been the subject of many studies by sociologists, economists, political scientists, and other thinkers.[2] The broadest attempt at stating the problem and proposing solutions was made at the United Nations conference on the Application of Science and Technology for the Benefit of the Less Developed Areas held in Geneva in 1963; the eight volumes of the report of the conference, *Science and Technology for Development,* deal severally with the changing world, natural resources, agriculture, industry, living conditions, education, and planning. The basic problem is how best to help populations that are largely dependent on subsistence farming, illiterate, and threatened by overpopulation to acquire the capital savings by which alone a nation can hope to take off into self-sustained economic growth (chapter 1). Importing food to stave off famine is clearly no long-term solution, since it provides no incentive for improvements. Programs to increase local agricultural yield are better in that regard, but even then no savings are possible if the extra yield is absorbed by a growing population. A nation whose food supply grows at a rate of 2 percent and whose population grows by 3 percent a year is obviously headed for catastrophe. Industrialization pure and simple, carried out with massive injections of foreign capital, is also no panacea. Politically it smacks of colonialism, especially if the donors try to keep some control over how their capital is used; and if they do not, the recipients are sometimes tempted to spend it on spectacular building or construction projects that are highly visible but of questionable worth in the economic "takeoff" process.

The optimum contribution that technologists can make is as part of a balanced program of agricultural reform, industrialization, and education. Here again, the enlightened engineer sees his role in a broader context of overall social change. He knows that the nations of the third world (the countries outside both the Western and the communist orbits) have their own cultural values and that these values are threatened by the indiscriminate imposition of a heedless technology that would, as the saying goes, make the whole world look like a California suburb.

The technologist's increasing awareness of the need for a broad view is evident from various sources. Most encouraging among such trends are the progressive attitudes of those concerned with the general problem of development and with the particular problem of the transfer of technology to developing countries. At an international conference on the Interdisciplinary Aspects of the Application of Engineering Technology to the Industrialization of Developing Countries, organized by the University of Pittsburgh's School of Engineering, a recurrent theme was the extent to which the success of industrialization depended not only on natural, financial, educational, and human resources, but also on such nontechnical considerations as the social, cultural, legal, and political environments. These relationships are delineated in considerable detail in the record of the meeting, *Industrialization and Development*.[3] At an earlier conference, devoted exclusively to the transfer of technology, this concern was expressed as follows by the Yale economist Neil Wolverton Chamberlain:

> The advancement of technology through transfer cannot be viewed as possessing some autonomous value, irrespective of the goals and values—confused and divergent though they be—that are built into the fabric of the receiving culture. The transfer of technology cannot be approached only with consideration of whether it is technically feasible; it must also be appraised in the light of its social acceptability. This raises further questions as to the extent to which both contributing and recipient countries should seek to cultivate receptivity to new techniques that will change a way of life, or the extent to which they should themselves be guided by considerations of social values that are found in the beneficiary country. And if the latter, then additional questions intrude: Which values, those of a traditional and culture-bound segment of the society, or those of its more innovative and opportunity-conscious members?[4]

This concern is not entirely new, though it used to worry none but anthropologists until aid to developing countries began to play a large role on the world's stage, as it has been since the middle of the 20th century. It was to an anthropologist, Margaret Mead, that the World Federation for Mental Health turned in 1954 to compile a manual for all those concerned with purposive technical changes, to show how they can be introduced with the least hazard to the mental health of the recipients. The result was the pioneering study sponsored by the United Nations Educational, Scientific and Cultural Organization (UNESCO), *Cultural Patterns and Technical Change* (1955), which sought to answer the question, "how can technical change be introduced with such regard for the culture pattern that human values are preserved?"[5] As a result of this and other studies, the realization has dawned that it would be naive, if not downright wicked, to pursue the great goal of the United Nations to promote "higher standards of living, full employment, and conditions of economic and social progress and development" (Article 55 of the 1945 U.N. Charter) by material means alone. More recent formulations of international aims sound a slightly different note. The charter of a U.N. agency founded in 1966 and concerned with industrial development, the United Nations Industrial Development Organization (UNIDO), while affirming that "the industrialization of developing countries is essential for their economic and social development and for the expansion and diversification of their trade," sees as one of the new organization's main purposes the contribution it can make "to the most effective application in the developing countries of modern industrial production, taking into account the experience of States with different social and economic systems." The last clause manifests a respect for local conditions and traditions that certainly represents a new departure.

The USA has led the world in providing aid to developing countries, not only in funds and goods dispersed but also in services, notably technical education.[6] Critics well endowed with cynicism have sought the motives for this unprecedented largesse in the economic and political advantages that supposedly accrue to the USA from its generosity, a viewpoint partially supported by the arguments sometimes advanced in the U.S. Congress when annual appropriations running into billions of dollars are sought to fund the various programs. It is unlikely that Western Europe, whose recovery from the war of 1939–1945 was speeded by American aid (the European Recovery Program initiated by the U.S. Secretary of State, General George Catlett Marshall, 1880–1959), would have been *more* anti-American without such aid, or that American agriculture would have sunk into destitution without the indirect subsidies represented by the Food for Peace Program (the disposal of surplus commodities). The actual explanation, that American aid has been generously given largely for humanitarian reasons, is so simple that anyone advancing it is thought to be practically light-headed.

Not all aspects of foreign aid proceed smoothly, and sometimes there are unexpected effects. One example is the so-called brain drain—the tendency of highly developed countries to attract specialists from less developed countries by better employment opportunities and working conditions. A shortage of specialists such as physicians or scientists in Canada or the USA attracts candidates from the United Kingdom; the resulting openings in the UK are filled by Indians or Pakistanis. Particularly vexing to the depleted countries is the emigration of graduate specialists who come to developed countries for postgraduate training and remain there. A study made by the author of the several hundred foreign students who received advanced degrees at the University of California's College of Engineering in Berkeley over a 12-year period, *Exporting Technical Education*, showed that nearly one-third of those that received masters' degrees and nearly two-thirds of those who attained the doctorate were still USA residents at the time of the survey, though some were planning to return ultimately.[7] (Only 2 percent came from Southeast Asia and Black Africa.)

The brain drain is not entirely a matter of differentials in professional opportunities, even though opportunity plays an important part: during recessions in the USA, the brain drain is actually reversed for some countries. Even in the above sample, few Latin American or French engineering students failed to return home, probably because of mutual prejudices, relative social standing in the home country, and other cultural considerations. Attempts to solve the problem of the brain drain have not been uniformly successful, although much can be done by administrative measures such as setting a national limit on the number of students admitted. Another approach is to bring the education to the students rather than the other way around, as India has done by letting America, Britain, Germany, and the Soviet Union each help set up an Institute of Technology in India; but even that is no solution if the new graduates remain unemployed and must seek engineering jobs abroad. More effective is a formal agreement under which a nation sends students to another country for advanced education (rather than letting them try their own luck) with the clear understanding that they are then to return home for at least several years. Among the reasons for a 95-percent return rate of the French engineering students in the above sample may be that they were almost all sent over for a limited period under French government auspices.[8]

In summary, the search for peace through international understanding presents a challenge that can be met in part by technology through cooperation, mutual aid, and education, provided the application of technology is not bedevilled by a narrow view of its capabilities and effects, and that its carriers respect the values, or tastes, or preferences of all peoples.

OVERPOPULATION *is everybody's baby.* The uncontrolled rise in the number of inhabitants of a region beyond the level it can support derives more than any other world problem from the advances of technology and is less susceptible to technical solutions than most problems. Better health and lower mortality result from improvements in public health measures, discoveries in pharmacology, better medical care, and other technical advances. Few would press for a return to the greater infant mortality and lower adult life expectancy of the "good old days." Yet better health means more conceptions, more pregnancies carried to term, more babies surviving, more young couples, and greater longevity—all factors leading to population increase.

Technology has been of some help through contraception, euphemistically called "family planning." But the whole array of means and techniques—mechanical devices, drugs, abortion, sterilization, clinics, modern methods of reaching semiliterate populations (for example, transistor radios)—is not altogether effective as a way of controlling population growth, as has been shown in detail by the demographer Kingsley Davis.[9] Family planning is not societal control. Such control might include encouragement of later marriages; income-tax policies favoring working wives, the unmarried, and small families; discouragement of large families by a partial reversal of current policies relating to welfare and public housing; and relaxation of rules regarding abortion and sterilization. Such measures are harsh even when no direct coercion is used; but nothing less will do very much good and the alternative—unchecked population growth—is sure to lead to horrors besides which the disadvantages of purposive government planning pale to insignificance. Technology can do very little except perhaps indirectly, by the industrialization of predominantly agrarian societies (a process that is known to reduce population growth significantly, though partially through undesirable effects, such as urban congestion), and by monitoring and helping to enforce whatever means of social control are attempted (for instance, through data acquisition and processing).

The problem of overpopulation has become the subject of much public debate, some of it a bit hysterical in tone.[10] The subject has a long history, going back at least to the work of Thomas Robert Malthus (1766–1834), who asserted that the growth of population would always outstrip that of production, since the former increased (if unchecked) in geometrical progression, whereas the latter could only grow arithmetically. Malthus supported his thesis with much qualitative evidence, but more recent scholars decry his data as inadequate and his logic and premises as faulty, although they recognize the historical importance of his theories. On the subject of population models and other aspects of "ecosystems," the pioneer study is that of Vito Volterra (1860–1940), the great Italian mathematician whose work in calculus led him to investigate the interdependence of animal species.[11]

Thoughtful writers have argued that overpopulation is one of a class of problems which, though arising in part from technological developments, have no purely technical solutions. (Another is the problem of decreasing national security in the face of increasing power to wage nuclear warfare, according to a pair of distinguished academics who had served as high officials of the U.S. Department of Defense.)[12] The best-known exposition of this argument is the article by Garrett Hardin, "The Tragedy of the Commons," based on the telling analogy of the pasturage set aside for common use by English villages without protection against overgrazing, underfertilization, and erosion.[13] The positive utility, to each herdsman, of adding one more animal to his herd far outweighs the negative utility (again to himself) of overgrazing, which is shared by all herdsmen. "Each man," says Hardin, "is locked into a system that compels him to increase his herd without a limit—in a world that is limited." His solution is bound to be controversial: abandonment of the freedom to breed, through "mutual coercion mutually agreed upon." No purely technical solution can rescue us from the misery of overpopulation. In summary, the challenge lies in recognizing that and concentrating our efforts on changing the ideology of the status quo instead.

LEISURE *is defined as "the state of being free of everyday necessity"* by the man who has made the most penetrating study of it, the American political scientist Sebastian de Grazia in his book *Of Time, Work, and Leisure*.[14] He carefully distinguishes leisure from "free time," the time off or time not related to work; leisure is a state, a condition in which time plays no part. How to turn free time into leisure is one of the problems of industrial society. According to De Grazia, free time has not grown much with the proliferation of technology, even though work time has significantly decreased in the past century; many Americans engage in "moonlighting" (a second job) and much supposedly free time is nowadays taken up with travel to and from the job, grooming and otherwise getting ready for it, shopping, maintaining the home, and other do-it-yourself projects.

Perhaps the most provocative work on the subject is the book by the British engineer and Nobel Prizewinner Dennis Gabor, *Inventing the Future*.[15] He feels that the age of leisure is almost upon us and cautions that in facing it without psychological preparation, our civilization risks a danger of self-destruction no smaller than that inherent in nuclear war or overpopulation. Only irrational waste keeps the common man in the industrial societies from having even more free time. Some of the waste is of the sort so wittily described by *Parkinson's Law* (1957), according to which work expands to occupy all the time available for it; any office worker worth his salt can create work for other office workers, without necessarily being aware that much of it is redundant and wasteful.[16] (A more optimistic view sees Parkinsonianism as society's defence against the lack of employment that automation would bring.) But even with waste we shall soon be at our wits' ends in dealing with the triple threat of nuclear extinction, overpopulation, and boredom. Gabor's prescription is for social science to copy technology and engage in something akin to systematic invention, accompanied by a no less systematic effort to obtain society's approval for innovation:

> The future cannot be predicted, but futures can be invented. It was man's ability to invent which has made human society what it is. The mental processes of invention are still mysterious. They are rational, but not logical, that is to say not deductive. The first step of the technological or social inventor is to visualize, by an act of imagination, a thing or a state of things which does not yet exist, and which to him appears in some way desirable. He can then start rationally arguing backwards from the invention, and forward from the means at his disposal, until a way is found from one to the other. There is no invention if the goal is not attainable by known means, but this cannot be known beforehand. The goal of the technological inventor is attainable if it is physically feasible,

but for the realization he will be dependent, just like the social inventor, on human consent. The difference is that while in the past many technological inventors failed tragically by not being able to obtain consent, this is today not only easy but often far too easy. For the social inventor, on the other hand, the engineering of human consent is the most essential and the most difficult step, and I do not think that this has become more easy in democracies where the masses must be persuaded, instead of perhaps one enlightened monarch.[17]

In summary, technology's role in social engineering thus appears to be that of a model, not only in the methods of innovation, but also in shifting the seeking of consent from emotional argument to rational persuasion.

TECHNOLOGY ASSESSMENT *is the evaluation of the adverse along with the beneficial effects of technological innovations.* As originally conceived in legislative circles in the USA, it was to be a method of obtaining information on the basis of which technological development could be stimulated, directed, and if necessary restrained, first in the public and then in the private sectors.[18] This idea contains the germs of several others. Implicit in the approach is the process of allocating resources and setting national priorities—a problem that free economies have historically encountered in times of crisis, such as a major war or depression.[19] Attention to this problem thus represents recognition that heedless technological development could create crises of similar magnitude. Beyond that, one is also led to seek more benign alternatives to existing technology, of "taming" technology, which is a still more complex matter.[20] Moreover, short of war, the strategies of entire nations have come to depend on technological developments. Thoughtful Europeans see the American challenge in the fields of technology and business management, not in the military sphere;[21] and the same holds for Asians and Japan. The relationship of countries such as France, Sweden, or Japan with more powerful nations is governed by dispassionate technological forces, a trend whose importance is still on the increase, according to an analysis made by the American political scientist Robert Gilpin:

Economic and technological considerations will shape the ways in which political interests and conflicts seek their expression and work themselves out. In a world where nuclear weaponry has inhibited the use of military power and where social and economic demands play an inordinate role in political life, the choice, success, or failure of a nation's technological strategy will influence in large measure its place in the international pecking order and its capacity to solve its domestic problems.[22]

Even a simple listing of "goods" and "bads" along the lines mentioned previously (chapter 7) is not really simple. The potential advantages of a proposed new departure—say, a supersonic transport aircraft—are usually obvious and are lovingly enumerated by its proponents, led by those who would directly profit by it. Opponents in turn enumerate the disadvantages. The opponents seldom include those who would be personally disadvantaged or even inconvenienced. More often, opposition comes from idealists working in what they conceive to be the public interest (although they may have other motives as well, such as envy and individual political or financial advantage). One of the requirements of a satisfactory system of technology assessment would thus seem to be more active involvement of those "whom it may concern." The term *participatory technology* has been coined for this concept.[23]

A related problem is that of advocacy in technology assessment. When the individual interests of two groups are at odds, more often than not they find themselves in an adversary position, like prosecution and defense in an Anglo-Saxon court.[24] Often it is not difficult for both to find technical experts to support the opposing viewpoints. If the adversaries are a government agency and a citizen group—say, the U.S. Bureau of Reclamation (in charge of developing flood and erosion control) and the Sierra Club (devoted to conserving wilderness areas)—the clash of experts quickly acquires political overtones, so that it becomes increasingly difficult for the lay observer to sort out technical from nontechnical issues or even to allow for bias and institutional advantage.[25] Yet the old lay prescription—when experts disagree, suspend judgment—is no longer of much use, since mere inaction becomes action (in a negative sense) in the face of inexorable and irreversible change.

There are also problems of a practical nature. In the USA, for example, many government agencies already seek to regulate the advance of technology, for instance in aviation, atomic energy, communications, food processing, and similar fields. Such agencies tend to be won over by the very industries they are meant to regulate, since many of the specialists on whose expertise the agencies depend are either recruited from these industries or look to them for future employment. Moreover, some agencies, such as the U.S. Atomic Energy Commission, are driven to bureaucratic schizophrenia by a dual assignment: to promote a technology and to regulate it at the same time. Nor are there formal provisions for integrating the assessments made by various agencies, each with its own proper interests and spheres of influence. But the greatest problem is that no real basis exists on which anyone could compute what constitutes the public interest. Even the most sophisticated method that has been applied to institutional decision making, *cost-benefit analysis,*[26] may not adequately concern itself with deleterious secondary consequences (what economists call "external diseconomies") of technological decisions.

The idea that external costs and benefits must be taken into account in computing economic efficiency goes back at least as far as the 1920s, beginning with the work of Arthur Pigou.[27] This pioneering concept has been carried farther by others, notably the American econometrician Kenneth J. Arrow. Pigou's successors pointed out long ago what now appears self-evident, that a sure way toward inefficiency is to allow private producers to engage in activities that impose costs they need not pay, or in activities that confer benefits for which no payment is received; for efficient allocation of resources there must be what we should call feedback—that is, external costs and benefits must be attributed to the companies that produce them and reflected in the prices of the goods and services they supply.

The concepts of cost-benefit analysis and of the social-welfare function in politics and economics are reckoned among the major advances in social science.[28] With regard to technology assessment, they are just a beginning. Further advances in the social sciences (law, economics, sociology) are urgently needed that can underlie a definition of the public interest extending beyond the notions of the 19th-century utilitarians on general welfare, to the needs of societies that have been through the Second Industrial Revolution.

Recognition that technology assessment is a political as well as a scientific process and should not be invested in a single authority came quite early. The very first panel convened by the National Academy of Sciences in response to a 1968 request from the U.S. Congress came to this conclusion.

> The panel would emphatically oppose any scheme that would empower an agency to decide, on behalf of something called "society" or "the environment," which technological developments will be permitted and which prohibited. Selections among alternative technologies require that choices be made among competing and conflicting interests and values. To the

extent that those choices are made and enforced collectively rather than individually, they are essentially political in character and must therefore be the responsibility of the politically responsive branches of government and of those publicly accountable bodies that are specifically entrusted with regulatory responsibilities in narrowly circumscribed areas. The making of such choices is, in principle, indistinguishable from the resolution of the many other conflicts that beset society. To entrust the resolution of all those conflicts to a single, all-encompassing authority would be incompatible with representative government.[29]

Therefore, the panel recommended that for maximum influence, any proposed mechanism should be placed close to the center of the political establishment but constrained "to study and to recommend but not to act." For the rest, the panel made few specific recommendations, noting ruefully that "the concept of improved technology assessment is by no means a unitary one; it suggests different things to different people."[30] The following were some of the notions thought to be of preeminent importance by various groups:

1) Preservation and enhancement of environmental quality: evaluation of the impact of technological changes on the environment and of the growth of technology on environmental objectives.

2) Measurement of social change: use of new tools to monitor the effect of technology on society and improvement of the quality of feedback from social to technological developments.

3) Greater foresight and planning: forecasting the probable consequences of particular technological developments to provide an "early warning system" of the potentialities and dangers of incipient technological developments.

4) Improving the allocation of public resources: identification of possible uses of government-sponsored technologies and use of the process of allocation of technological priorities as a catalyst for new social and political departures.

5) Better program and policy evaluation: extension of the Planning-Programming-Budgeting System toward "developing more precise definitions of program objectives as they relate to national goals and priorities; more specific and unbiased criteria for assessing program potentiality and performance in cost-benefit terms; and more successful ways of modifying old programs or proposing new ones with the help of such analytic devices."[31]

Although technology assessment is still in a formative stage, so that it is difficult to estimate which of these notions will come to be included and in what sequence, a consensus is beginning to emerge.

The report of a similar panel established independently by a sister body, the National Academy of Engineering, parallels some of the above concepts. As it is written by engineers rather than scientists, the NAE report is a little more down to earth than the NAS report: it includes three examples of assessments attempted in the recent past (teaching aids, subsonic aircraft noise, and multiphasic health screening); makes specific "how to" recommendations on optimizing the usefulness of technology assessment, broadening the concept, and deriving maximum benefits from it; and specifies the roles that it might be expected to play:

1) *Clarifying* the nature of existing social problems as they are influenced by technology, possibly with indications of legislation needed to achieve satisfactory control.
2) *Providing* insights into future problems, to make possible the establishment of long-term priorities and to provide guidance for the allocation of national resources.
3) *Stimulating* the private and public sectors of our society to take those courses of action for the development of new technology that are most socially desirable. Such actions may be creative or defensive. Creative actions would be those that follow from the awareness of new opportunities for social development; defensive actions would be those involving restrictions on the use of technological developments.
4) *Educating* the public and the government about the short-term and long-term effects of the range of alternative solutions to current problems.[32]

It may well be that what is most urgently needed is an assessment of technology assessment, some hard work to arrive at nationally and internationally acceptable criteria against which the "goods" and "bads" of technological developments can be measured. The problem has puzzled lawyers, economists, and philosophers for generations. "Almost without exception," says the NAS report, "technological developments will affect some people or interests adversely and others beneficially, and there simply is no agreed-upon algebra by which one can neatly subtract the pains from the pleasures in order to arrive at a net index of social desirability."[33] Nor can one fall back on the "greatest good for the greatest number" goal of 19th-century utilitarianism: in an even moderately complex society, it is impossible to arrive at an optimum solution merely by piling up the individual preferences of various groups on top of one another.[34] A better hope is that methods and tools derived from modern technology, such as system theory and computers, when combined with the best brains that economics and other social sciences can supply, will provide the answer.

In summary, technology assessment looms as an urgent task for all nations; unrestricted laissez faire is no more viable in large-scale technological development than in big business, though care must be taken that in bending technology to the purposes of society, individual initiative and private enterprise are not inadvertently pushed out of existence. In putting technology assessment into practice, both intuition and analysis must be brought to bear. In developing the requisite analytical techniques, we shall surely find that technology, the source of the problem, will once again prove to contain within itself the germs of a solution compatible with the betterment of man's lot and dignity.

ALIENATION, *the resentment and disaffection experienced by the member of an unresponsive society* that subjects him to forces beyond his control and understanding, is the final challenge we shall consider. It has been a challenge to industrialized societies for two centuries. The problem has reappeared in our time as a result of the Second Industrial Revolution, most noticeably in the attitudes of college students and adolescents in highly industrialized countries, but more potently in the dissatisfactions of the much larger sector of the population usually described as disadvantaged—handicapped by poverty, prejudice, and lack of the skills demanded by mechanized industry.[35]

The two groups share a fear of a future based on a view of a society that will have no use for them. That view has been supported by the Cassandra cries of those who see automation as leading to chronic unemployment, and to increasing alienation among the few who manage to keep jobs. Neither outcome is inevitable. Mechanization does away with some jobs, creates others. A well-known example is the dial telephone. Before dialing was introduced, operators at local telephone exchanges manually connected the caller with the party called. It is estimated that if that system were in operation in industrialized nations today, half the adult population would be needed to man switchboards. Yet replacement of operators by automatic equipment and allied advances have led to a tremendous expansion of the telephone industry, the employment of tens of thousands, and work that is less routine (and thus presumably less alienating). The trend toward the employment of more office ("white-collar") workers and fewer production ("blue-collar") workers is well established in industrialized societies; so is the concurrent trend toward a higher proportion of professionals in the work force. These massive shifts resulting from the Second Industrial Revolution are comparable to the changes that have gone with all major reorganizations of society. The reorganizations themselves are well-nigh inevitable: nothing is stronger than an idea whose time has come. We look to the sciences of the individual and of the collective mind—to psychology and sociology—to show us how the requisite changes might be accomplished with least human cost.

The existence of individual costs was highlighted during the 1960s by the so-called student revolt, or New Left movement, in the advanced Western countries and Japan. Directed against technology in general and the USA (as its most successful practitioner) in particular, college student movements in several countries struck poses reminiscent of the Luddites (chapter 1), though addressing their resentments in the first instance against their universities, not their parents or political power centers. The true believers were seldom students of medicine, engineering, architecture, the natural sciences, or other "useful" professions; they were more likely to be students of the humanities and social sciences who could not foresee any roles for themselves in what they saw as a highly technological and allegedly materialistic society. They feared they had become what psychologist Bruno Bettelheim called "obsolete youth," young people who were often well endowed materially and spiritually, but who were thoroughly demoralized by the devastating thought that they, who should be the future leaders if there were any justice, might turn out to be unwanted and unnecessary—faceless numbers in an anonymous array run by technocrats.[36] (A typical protest slogan of the "Free Speech" movement at Berkeley in 1964 parodied the instructions that go with punched computer cards: "I am a human being: Do not fold, spindle, or mutilate.")

With time, the realization dawned that we were not necessarily headed for a society in which technologists would form the only—or even the most—important sector. To begin with, increasing productivity is reducing the numerical predominance of production workers, just as the improved effectiveness of modern agriculture served to reduce the numbers of workers engaged in that. Whereas industrial societies see production as their central activity today, they may come to be organized around quite a different core tomorrow—education, for instance, and social services generally. It is not difficult to visualize a society with nearly half its members engaged in teaching the other half and with education expanded from an activity of the very young to a lifelong pursuit. Similarly, it is possible to foresee a society that devotes a much larger portion of its efforts to social services and to intellectual and artistic endeavors. In such a society, with most of its members freed from routine and deadening tasks, those devoting their lives to the social sciences and to the arts and humanities will be in great demand; and a member's usefulness will not be judged solely by his contributions to industrial productivity.

Nor is there any obvious reason that the society of the future need be organized along the sterile and unsatisfactory lines predicted by the pessimists. Alternatives abound. One is the *active society* proposed by sociologist Amitai Etzioni, who foresees a society made up of citizens consciously committed to specific societal goals and in control of the power needed for their realization.[37] One of the major obstacles—and major sources of alienation—is inauthenticity, the appearance of responsiveness in a basically unresponsive environment, notably the inauthenticity of many political processes, even in democracies. A man is alienated, says Etzioni, if he has been reduced to an object or has become entangled in unresponsive projects that involve treating others and himself as things and that prevent responsiveness to basic human needs. The remedy is the simultaneous strengthening of free society and unremitting activity—simultaneous because we seem to face something akin to a closed circle: action can only be effective in a free society, and society can only be free if there is continued action. Etzioni balefully distinguishes three types of citizens:

> In the inauthentic society, the majority of the members are caught in the typical cleavage between their private selves and public roles and manage by treating their neuroses with drugs, alcohol, professional counseling, and the like, thus reinforcing the inauthenticity of the society which caused their malaise. There is a minority of retreatists who ignore their public roles and build lives around their private selves. While these people are more authentic and, potentially, carriers of societal change, they have little societal effect. Finally, there are those who evolve new public selves which they collectivize and make the basis of their societal action. In these lies the hope for an initiation of the transformation of the inauthentic society. They are the active ones.

And they, one may add, are the ones who will create the democratic society of tomorrow—not by shouting simplistic slogans, not by railing against the advances of technology, not by dropping out—but by tuning in on the real problems of mankind and doing something about their solution.

Notes

*Notes to Chapter 1**

1. Benjamin Franklin defined man as the tool-making animal, a concept that is also characteristic of the work of the British anthropologist Kenneth P. Oakley, author of *Man the Tool-Maker* (London: British Museum, 1949[pb]). More recent findings suggest that the use of weapons antedated tools and characterized man's prehuman ancestor *Australopithecus africanus*, a notion that some writers think may come to have a profound effect on contemporary thought; see, for instance, Robert Ardrey, *African Genesis* (New York: Atheneum, 1961[pb]).
2. Barbara Ward, *The Rich Nations and the Poor Nations* (New York: W. W. Norton & Co., 1962[pb]).
3. Journals in English primarily devoted to the history of technology include *Journal of Industrial Archeology*, *Transaction of the Newcomen Society*, and *Technology and Culture*. The last is the official quarterly of the USA-based Society for the History of Technology (SHOT), whose secretary, Melvin Kranzberg, coedited (with Carroll W. Pursell, Jr.) the two-volume *Technology in Western Civilization* (New York: Oxford University Press, 1967). A more detailed collection is the five-volume *History of Technology* (London: Oxford University Press, 1954–1958), coedited by Charles Singer, E. J. Holmyard, A. R. Hall, and T. I. Williams. Another European contribution (translated from the French) is the two-volume *A History of Technology and Invention: Progress Through the Ages*, edited by Maurice Daumas (New York: Crown Publishers, 1969). A classical contribution to the field is Lewis Mumford's *Technics and Civilization* (New

* Original publication dates of books are given; a superscript "pb" signifies that a more recent paperback edition is also available.

York: Harcourt, Brace and Co., 1934pb); others are the histories of engineering by James Kip Finch, *Engineering and Western Civilization* (New York: McGraw-Hill, 1951), and of invention by Otis T. Mason, *The Origins of Invention* (Cambridge, Mass.: MIT Press, 1966pb). Universities in several countries offer studies leading to the doctorate in the history of technology.

4. An excellent survey of the Industrial Revolution era in Britain and France is E. J. Hobsbawm's *The Age of Revolution: 1789–1848* (London: George Weidenfeld and Nicholson; New York: World Publishing, 1962bp). An important contemporary source is the collection of biographies by Samuel Smiles, *Lives of the Engineers* (London, 1857 and 1904bp).

5. The generalizations about religious attitudes admit of exceptions: many useful inventions have been made by individual Catholics, and Europe depended on the East for important innovations over long periods at a stretch. On the other hand, there is considerable evidence that the remarkably fruitful period of medieval Islamic science and technology came to an end in the 12th century as a result of the ascendancy of a particular theological viewpoint.

6. The principals in the development of textile technology were John Kay (1704–1764), Edmund Cartwright (1743–1823), Joseph Marie Jacquard (1752–1834), John Wyatt (b. 1700), Lewis Paul (d. 1759), James Hargreaves (d. 1778), Sir Richard Arkwright (1732–1792), and Samuel Crompton (1753–1827).

7. Hobsbawm (see note 4 above) says that Mohammed Ali's efforts were undermined by the Anglo-Turkish Convention of 1838, which opened Egypt to foreign traders, and by the reduction of his army dictated by Britain and France after Egypt's defeat over the period 1839–1841.

8. According to another account, the name Luddites is derived from "the tradition in the hosiery industry that a certain Ned Luddam, a Leicester stockinger's apprentice, when reprimanded on one occasion, and ordered to square his frames, had lost his temper and taken up a hammer and beaten the offending frames into pieces." Cited from Blackner's *History of Nottingham* in F. O. Darvall, *Popular Disturbances and Public Order in Regency England* (London: Oxford University Press, 1934), p. 1. Darvall sees Luddism mainly as an "attempt to apply pressure to certain employers, and to force them to grant the body of employees various concessions which they were demanding, and which they had been unable to obtain by pacific means" (*ibid.*, p. 3).

9. Utilitarianism, originally a concept in ethics, holds that the greatest happiness of the greatest number should be the principal criterion of morality. (See, for instance, J. S. Mill, *Utilitarianism*, 1863.) The concept has undergone considerable modification, notably at the hands of the English philosopher George Edward Moore (1873–1958) and more recently by population experts who point out the catastrophic consequences of its untrammeled application in overpopulation.

10. Admittedly, abolition of slavery in the USA might have taken much longer without the war, especially since industrialization progressed very slowly in the southern states until quite recently.

Notes to Chapter 2

1. Prince Albert's dream of instituting a German-type "Technische Hochschule" (like the one at Charlottenburg near Berlin) in London remained unfulfilled, partly because of his premature death in 1861. The institutions established

on the South Kensington site in London, a site bought with the profits from the Great Exhibition of 1851, were mainly museums and only secondarily schools of science and technology. The organization of the schools into the top-ranking Imperial College of Science and Technology, with status as a college in the University of London, did not take place until the 20th century.

2. In 1965, a group of historians of technology and social scientists on both sides of the Iron Curtain formed the International Cooperation in the History of Technology Committee (ICOHTEC), with headquarters in Paris and support from UNESCO, for the purpose of studying the transfer of technology, among other questions.

3. Lord Blackett, the Nobel-Prize-winning physicist who helped develop operations research and who became president of the Royal Society in 1965, wrote in 1935: "The Second Law of Thermodynamics arose from the attempt to make steam engines more efficient. . . . Today this Second Law of Thermodynamics appears one of the most far-reaching of all physical laws. . . . So the most abstract and general of laws arose from the study of that most concrete of objects, the steam engine."

4. Rankine became professor of engineering at the University of Glasgow in 1855. The universities of Scotland, like their society more egalitarian in outlook and composition of the student body than their English counterparts, were the first in Britain to award university status to engineering studies (1840).

5. A splendid essay on Bessemer's invention appears in E. E. Morison's *Men, Machines, and Modern Times* (Cambridge, Mass.: MIT Press, 1966pb).

6. For a "biography" of the Great Eastern, see James Dugan, *The Great Iron Ship* (London: Hamilton, 1953), which originally appeared in *The New Yorker* magazine.

7. The first one-cent newspaper was the New York *Sun*, established in 1833 by Benjamin Henry Day (1810–1889). The penny post was evolved in 1837 by Sir Rowland Hill (1795–1879) and adopted in 1839.

8. One news agency antedated the telegraph: Agence Havas (now Agence France-Presse) was founded in Paris in 1835.

9. A major work tracing the history of mass communications and its effect on people is a compendium of 29 articles collected by Lewis Dexter and David White, *People, Society, and Mass Communications* (New York: Free Press, 1964), which examines the nature of modern society as influenced by mass communications.

10. Early Russian work on radiotelegraphy by Aleksandr Stepanovich Popov (1859–1905) has become the subject of extravagant Soviet claims about priority in the invention of radio; for a critical examination of these claims, see C. Susskind, *Popov and the Beginnings of Radiotelegraphy* (San Francisco: San Francisco Press, 1962 and 1973).

11. The first Roman aqueduct was built by Appius Claudius Caecus in 312 B.C.; the Hagia Sophia, originally a Christian church, was built by the Emperor Justinian at Constantinople (now Istanbul) during A.D. 532–537; the Great Wall of China dates mainly from the Ming Dynasty (A.D. 1368–1644).

12. A role similar to Ford's was played in Europe by the French auto manufacturer Louis Renault (1877–1944) and the Czech shoe manufacturer Tomáš Baťa (1876–1932), whose factory community at Zlín was taken over by the state in 1949 and renamed Gottwaldov, after the first communist president of Czechoslovakia, Klement Gottwald (1896–1953).

13. A program of social insurance was introduced in Germany in 1883 by the first chancellor, Otto von Bismarck (1815–1898), partly to weaken the appeal of the socialists and partly to advance his policy of economic nationalism, which he formulated to replace laissez faire.

14. Outstanding among Taylor's successors were Henry Lawrence Gantt (1861–1919) and the husband-and-wife team of Frank Bunker Gilbreth (1868–1924) and Lillian Moller Gilbreth (1878–1972), the parents in the 1949 bestseller *Cheaper by the Dozen.*

15. The Hawthorne experiments, which were carried out during 1927–1932, were reported by Elton Mayo, *The Human Problems of an Industrial Civilization* (New York: Macmillan, 1933pb); see also E. H. Spicer, ed., *Human Problems in Technological Change* (New York: Russell Sage Foundation, 1952). In the USSR, the movement to speed production by efficient working techniques began in 1935 and was named after Aleksei Grigorevich Stakhanov, a coal miner who led a crew to a sevenfold increase in output.

16. For an early study concerned mainly with "blue collar" (production) workers, see A. R. Heron, *Why Men Work* (Stanford, Cal.: Standford University Press, 1948).

17. Crozet had served as an artillery officer under Napoleon before emigrating to the USA; after teaching at West Point from 1816 to 1823, he became a successful civil engineer in Virginia and in 1860 served as a member of a board to revise the program of instruction at West Point. Of the 59 books on "architecture, bridges, canals, perspective, and topography" listed in the earliest printed catalog of books in the West Point library (1822), 45 are in French. This library, which two officers had been sent to Europe to acquire during 1815–1817, thus "served as a major vehicle for the introduction of French technology into American culture," wrote Sidney Forman, West Point librarian, 150 years later.

18. A detailed description of the obstacles encountered by engineering education on its way to college-level status, with emphasis on the British experience, is given by Sir Eric Ashby, *Technology and the Academics* (London: Macmillan, 1958pb).

Notes to Chapter 3

1. The actual inventor of aspirin was Felix Hoffmann (1868–1946), a young pharmacologist who had only just entered the Farbenfabriken Bayer firm in Elberfeld after a year of postdoctoral study in the chemistry department of the University of Munich. He is not to be confounded with August Wilhelm von Hofmann (1818–1892), the great German organic chemist who became head of the Royal College of Chemistry in London and taught Perkin. The principal component of aspirin, salicylic acid, also occurs naturally in many plants.

2. The classical work on this topic is S. Collum Gilfillan, *The Sociology of Invention* (Chicago: Follett Publishing Co., 1935pb) and his more recent *Supplement to the Sociology of Invention* (San Francisco: San Francisco Press, 1971).

3. The six countries in which radar was simultaneously developed: America, Britain, France, Germany, Italy, and Japan. The amazing story of its development during the decade 1935–1945 is the subject of a forthcoming study by the author, *The Birth of the Golden Cockerel: A History of the Development of Radar.*

Notes to Chapter 4

1. Norbert Wiener's autobiography appeared in two volumes, *Ex-Prodigy* and *I Am a Mathematician* (New York: Simon and Schuster, 1953 and 1956pb). His *Human Use of Human Beings* (Boston: Houghton Mifflin, 1950 and 1954pb) popularized some of the ideas implicit in the highly technical *Cybernetics* (Boston: Houghton Mifflin, 1948 and 1961pb). His last book, *God & Golem, Inc.* (Cambridge, Mass.: MIT Press, 1964pb), is subtitled: *A comment on certain points where cybernetics impinges on religion.*

2. John Robinson Pierce, author of *Symbols, Signals and Noise* (New York: Harper, 1961pb), one of America's top electronics engineers, also played an important part in the introduction of communications satellites, a development he described in *The Beginnings of Satellite Communications* (San Francisco: San Francisco Press, 1968). He is also a prolific author of science fiction, under the pseudonym J. J. Coupling.

3. Detection at the receiver is the recovery or extraction of information that was superposed on the transmitted pure ("carrier") signal at the transmitter. The process thus differs from rectification, although the same device may be used for both.

4. Of the many descriptive works about electronic computers, two are especially good: D. G. Fink, *Computers and the Human Mind* (New York: Doubleday, 1966), an original paperback introduction to artificial intelligence; and Jeremy Bernstein, *The Analytical Engine* (New York: Random House, 1964pb), which originally appeared in *The New Yorker* magazine.

5. Hollerith's name survives in a type of punched card known as "Hollerith cards." For his subsequent career, see the note by his daughter, Virginia Hollerith, "Biographical sketch of Herman Hollerith," *Isis* 62(1971):69–78.

6. Goethe's ballad *Der Zauberlehrling* first appeared in *Musen Almanach für das Jahr 1798* (Tübingen: J. G. Cotta, 1798). Both Offenbach's *Les contes d'Hoffmann* and Délibes' *Coppèlia* were based on the same story of Hoffmann, "Der Sandmann," which appeared in his collection *Nachstücke* in 1816. (Act I of Offenbach's opera is based on a portion of the tale; the ballet is "a humorous parody" of it.) Mary Godwin Shelley (1797–1851) wrote the classic horror tale about a German student who learns how to infuse life into matter and creates a monster which destroys him. (Frankenstein is the name of the creator, not of the monster.) The interest in mechanical automata derived from the ingenious pump and clockwork creations, usually in the form of animal or human figures, of 18th-century mechanicians such as Jacques de Vaucanson (1709–1782).

7. A widely known example is Arthur C. Clarke's *2001—Space Odyssey*, based on the 1969 film script he coauthored with Stanley Kubrick. Both authors disclaim as an unintentional coincidence the fact that HAL, the nickname of the computer that directs the spaceship, becomes IBM when each letter is shifted one position in the alphabet.

8. A. M. Turing's article, based on a paper in *Mind* 59(1950):433–460, is a chapter in J. R. Newman, ed., *The World of Mathematics* (New York: Simon and Shuster, 1956pb), vol. 4, pp. 2099–2123; the cited passage is on pp. 2109–2110. Newman also served as the editor of the Harper Modern Science series in which Pierce's 1961 book on information theory (see n. 2 above) was published.

9. Gabor's paper, which did not include a detailed treatment of noise, appeared in the *Journal of the Institution of Electrical Engineers* in London:

D. Gabor, "Theory of Communications," *JIEE* 93(1946):111 and 429. Shannon's work was first summarized in his company's *Bell System Technical Journal*: C. E. Shannon, "Mathematical Theory of Communications," *BSTJ* 27(1948):379–423 and 623–656, before it was published in book form (Urbana: University of Illinois Press, 1949pb).

Notes to Chapter 5

1. The section on "The New Food Revolution" is based on an article of that title by Gene Gregory in *The UNESCO Courier* (published 11 times a year by the United Nations Educational, Scientific, and Cultural Organization) for March 1969; this fine magazine's practice of making such materials available for adaptation is gratefully acknowledged. The quest for increased food production is the subject of many books; one well-told account is Paul de Kruif's *Hunger Fighters* (New York: Harcourt, Brace and Co., 1928pb). For the challenge presented by overpopulation, see chap. 8.

2. Research on aggression is at a relatively early stage and only incidental applications of technology have been made to it (for example, in the use of behavior-altering drugs and of brain-wave recorders). The situation is changing, perhaps owing partly to the attention that has been focused on the problem by a number of general-interest books about the animal roots of human behavior, such as Robert Ardrey, *Territorial Imperative* (New York: Atheneum, 1966pb); Konrad Lorenz, *On Aggression* (London: Methuen, 1967pb; originally published as *Das sogenannte Böse* [Vienna: Borotha-Schoeler Verlag, 1963]); and Desmond Morris, *The Naked Ape* (London: Jonathan Cape, 1967pb). But some scientists are increasingly skeptical of theories that human behavior is based on instincts; see, for instance, Leon Eisenberg, "The *human* nature of human nature," *Science* 176(1972):123–128.

3. Joseph (later Sir Joseph) Paxton was head gardener at Chatsworth, the Duke of Devonshire's huge estate, whose greenhouses doubtless gave him the idea (and prior experience) for the Crystal Palace. The subcontracting arrangements for the structural parts and the manner in which their dimensions and strengths were checked, highly suggestive of modern prefabrication methods, were a century ahead of general architectural practice.

4. The quotation is from a book by Walter Gropius, *The New Architecture and the Bauhaus* (London: Faber & Faber, 1935pb), p. 36. The original German version, *Die neue Architektur und das Bauhaus*, could not be published in Nazi Germany and did not in fact come out until 1965pb.

5. Among the slowest of the traditions of architecture to change has been the size of the work force required on a construction project, but even that is giving way to automation. In 1928, for instance, the world of architecture was shocked when a nearly completed 7-story reinforced-concrete building collapsed in Prague (the first in a series of accidents later traced to hasty work by contractors who were trying to beat a deadline before a new tax went into effect), burying over 100 construction workmen. Today, a smaller crew is visible in putting together a modern skyscraper.

6. Klüver's essay on Tinguely's "Homage to New York" appears in the catalog of a show organized by K. G. Pontus Hulten, *The Machine as Seen at the End*

of the Mechanical Age (New York: Museum of Modern Art, 1968), pp. 168–171; the cited portion appears on p. 171.

7. From an EAT advertisement in the *New York Times*, 12 November 1967, p. D 38.

8. Charles Frankel, "Education and telecommunications," *Journal of Higher Education* 41(1970):1–15; the cited portion is on p. 4. Professor Frankel, never an ivory-tower philosopher, put in a two-year stint as assistant secretary for educational and cultural affairs of the U.S. Department of State during one of the Johnson administration's abortive attempts to involve academics in making and carrying out science policy.

9. Anthony G. Oettinger with Sema Marks, *Run, Computer, Run: The Mythology of Educational Innovation* (Cambridge, Mass.: Harvard University Press, 1969). The title parodies the phrase familiar to millions of American first-grade readers: "Run, Spot, run!" Professor Oettinger, a specialist in linguistics and applied mathematics, correctly predicted in his own doctoral dissertation (in 1954) that a translating machine capable of presenting a well-written English version of a Russian text would remain a practical impossibility for decades, a conclusion with which a special committee of the National Academy of Sciences ruefully agreed twelve years later, after a government expenditure of nearly $20 million. (To put that sum into context, it may help to remember that not all machine-translation research—and the concomitant employment, development, and graduate training—was a total loss; and that the bill for the equally fruitless project of designing a nuclear-powered aircraft was $1500 million.) The principal contentions of *Run, Computer, Run* were first subjected to professional scrutiny in essay form, with discussion and author's reply, in "Educational Technology: New Myths and Old Realities," *Harvard Educational Review* 38(1968):697–755. The project was part of the IBM-endowed Harvard Program on Technology and Society, whose controversial annual reports in defense of technology proved to be unacceptable to left-wing critics—notably the widely circulated Fourth Annual Report (1967–1968), which received a devastating review in a special supplement of the *New York Review of Books* (John McDermott, "Technology: The Opiate of the Intellectuals," 31 July 1969).

10. U.S. House of Representatives, Committee on Education and Labor, *To Improve Learning: A Report to the President and the Congress of the United States by the Commission on Instructional Technology* (Washington, D.C.: U.S. Government Printing Office, 1970); National Academy of Engineering, *Educational Technology in Higher Education: The Promises and Limitations of ITV and CAI*, Report of the Instructional Technology Committee of the NAE Commission on Education (Washington, D.C., 1969); R. E. Levien, ed., *Computers in Instruction: Their Future in Higher Education*, Proceedings of a Conference Sponsored by National Science Foundation, Carnegie Commission on Higher Education, and Rand Corp. (Santa Monica, Cal.: Rand Corp., 1971); *The Application of Technology to Education*, special issue of *Journal of Engineering Education* 59 (February 1969); "Special Issue on Engineering Education," *Proceedings of the Institute of Electrical and Electronics Engineers* 59 (June 1971).

11. Oettinger, "Educational Technology," p. 708.

12. The concept of "two cultures" was enunciated by the British scientist-novelist C. P. Snow in his 1959 Rede Lecture at Cambridge University, *The Two Cultures and the Scientific Revolution* (London and New York: Cambridge Uni-

versity Press, 1959). A revised edition was published 5 years later: *The Two Cultures; and a Second Look* (London and New York: Cambridge University Press, 1964).

13. This portion of the text is based on the research topic report of A. L. Hammond, "Tools for Archeology: Aids to Studying the Past," *Science* 173(1971):510–511.

14. The radiocarbon dating technique was developed during the 1950s by Willard Frank Libby and his students at the University of Chicago, work for which he received the Nobel Prize in physics in 1960.

Notes to Chapter 6

1. The phrase, "the career open to talents," occurs in vol. 1, p. 103, of *Napoleon in Exile* (1822), by the Irish physician who attended the deposed emperor on St. Helena, Barry Edward O'Meara (1786–1836). The other saying is usually rendered in the original as *tout soldat français porte dans sa giberne le baton de marechal de France.* Much of Napoleon's appeal to the intellectuals of his day doubtless derived from his insistence on equality of opportunity; for instance, it is a major drive for the heroes of both of Stendhal's masterpieces, *The Red and the Black* (1831) and *The Charterhouse of Parma* (1839).

2. Veblen received the Ph.D. (in philosophy) from Yale in 1884 and also studied at Cornell and Chicago; he taught at Chicago, Stanford, and the University of Missouri. *The Engineers and the Price System* first appeared in 1919[pb] as a series of essays in *The Dial.* For an appraisal of the book in light of the Second Industrial Revolution, see J. M. Gould, *The Technical Elite* (New York: A. M. Kelley, 1966), which also contains good statistical data on various factors affecting industrially advanced societies.

3. Ogburn's ideas were summarized and elaborated in a book that he coauthored just before he died: F. R. Allen, H. Hart, D. C. Miller, W. F. Ogburn, and M. F. Nimkoff, *Technology and Social Change* (New York: Appleton-Century-Crofts, 1957). This topic has been carried further in many works, mainly by sociologists and anthropologists, a few of which are mentioned below. A well-known work is Jacob Schmookler, *Invention and Economic Growth* (Cambridge, Mass.: Harvard University Press, 1966). His ideas, notably on the role of patents in fostering invention, were hotly contested in a special issue of *Technology and Culture* (Summer 1960) devoted to that question, among others by S. C. Gilfillan, whose own *Sociology of Invention* (1935) and the subsequent *Supplement to the Sociology of Invention* (San Francisco: San Francisco Press, 1971) were mentioned in n. 2 of chap. 3. Other books include H. G. Barnett, *Innovation: The Basis of Cultural Change* (New York: McGraw-Hill, 1953[pb]); A. C. Crombie, ed., *Scientific Change: Historical Studies in the Intellectual, Social and Technical Conditions for Scientific Discovery and Technical Invention* (New York: Basic Books, 1963); F. A. Bond, ed., *Technological Change and Economic Growth* (Ann Arbor, Mich.: University of Michigan, 1965[pb]); and Robert Boguslaw, *The New Utopians: A Study of System Design and Social Change* (Englewood Cliffs, N.J.: Prentice-Hall, 1965[pb]).

4. A detailed history of technocracy appears in Henry Elsner, Jr., *The Technocrats: Prophets of Automation* (Syracuse, N.Y.: Syracuse University Press, 1967). An example of how the word's meaning has been enlarged is the title of "a

social history" by W. H. G. Armytage, *The Rise of the Technocrats* (London: Routledge and Kegan Paul, 1965).

5. James Burnham, *The Managerial Revolution* (New York: John Day, 1941[pb]), p. 203.

6. Orwell's critique of Burnham first appeared in the magazine *Polemic* (and as a London Socialist Book Centre reprint in 1946) and has been reprinted in various anthologies of Orwell's writings, for example, in *The Orwell Reader* (New York: Harcourt, Brace & Co., 1956), pp. 335–354.

7. John Maynard Keynes first earned public notice with his critique of the 1919 Versailles treaty, *Economic Consequences of the Peace* (1919). His chief work was *The General Theory of Employment, Interest, and Money* (1936[pb]). He was one of the architects (at Bretton Woods in 1944) of the World Bank (more exactly the United Nations' International Bank for Reconstruction and Development) and of the International Monetary Fund.

8. *The New Industrial State* was published in Boston by Houghton Mifflin Co. (1967[pb]). For criticism, see, for instance, the review by R. M. Solow, Galbraith's response, and Solow's rejoinder, in *Public Interest* 9(Fall 1967):100–119; see also the second edition of *The New Industrial State* (Boston: Houghton Mifflin Co., 1971[pb]). Among the points made by various economists who contest Galbraith's ideas are that professional managers are a long way from a total takeover—bankers and lawyers are still central to industrial decision making; that the motivation of managers is by no means as clear cut as Galbraith claims; that many obvious examples show that even an industrial combine with massive resources cannot manipulate demand with much certainty; that government intervention in big business was not introduced in the USA during the administration of J. F. Kennedy but much sooner, notably under F. D. Roosevelt but going back at least as far as the Sherman Antitrust Act (1890); and above all, that the market has *not* been supplanted by central planning in advanced countries—in fact, that there is virtually *no* central planning of large-scale industrial activity, since the market actually provides a substitute for planning.

9. President Eisenhower's remarks were made in his farewell address to the nation on 17 January 1961. They horrified his special assistant for science and technology, George Bogdan Kistiakowsky, who was trying to recruit specialists for government service and quickly disavowed his chief in a "footnote to history" (*Science* 133[1961]:355).

10. The similarities between "capitalist" and "socialist" industrial management have been noted by many observers; see, for instance, Reinhard Bendix, *Work and Authority in Industry* (New York: Harper and Row, 1956[pb]).

11. Relevant works of J. K. Galbraith, all published in Boston by Houghton Mifflin Co., include *American Capitalism: The Concept of Countervailing Power* (1952[pb]), *The Affluent Society* (1958[pb]), *The Liberal Hour* (1960[pb]), and *The New Industrial State* (1967[pb]). In addition to being Warburg Professor of Economics at Harvard, Galbraith has been active in public life, most notably as U.S. ambassador to India (1961–1963).

12. The original title of Spengler's work is even gloomier: *Der Untergang des Abendlandes* (that is, Fall of the Occident). Oswald Spengler (1880–1936) prophesied the conquest of the West by the yellow race. His advocacy of obedience to the state found ready response among Nazi ideologues, but he was discarded when he refused to support their racial theories.

13. Friedrich Georg Jünger is not to be confounded with his brother, the novelist Ernst Jünger, likewise a critic of technology.

14. Jacques Ellul is professor of history and contemporary sociology at the University of Bordeaux. He is also a prominent Catholic layman who has been active in the ecumenical movement. Five of his books have been translated into English and published in New York by Knopf: *The Technological Society* (1964pb), *Propaganda* (1965), *The Political Illusion* (1967), *A Critique of the New Commonplaces* (1968), and *Autopsy of a Revolution* (1971).

15. "The Technological Order," in *Technology and Culture* 3(Fall 1962): 394–421.

16. The term "big science," meaning scientific establishments with multimillion-dollar budgets drawn from public funds, is not Ellul's but was coined later. See, for instance, D. J. deSolla Price, *Little Science, Big Science* (New York: Columbia University Press, 1963pb); and A. M. Weinberg, *Reflections on Big Science* (Cambridge, Mass.: MIT Press, 1967pb).

17. This concept was first given its current meaning in W. H. Whyte's book *The Organization Man* (New York: Simon and Schuster, 1956pb).

18. Of the five French authors mentioned, two are available in English translation. Free Press has published Georges Friedmann's *The Anatomy of Work: Labor, Leisure, and the Implications of Automation* (1961pb) and *Industrial Society* (1955pb). Harper & Row is the publisher of several books by the Jesuit scientist-philosopher Pierre Teilhard de Chardin (1881–1955), notably *The Future of Man* (1964pb), a series of papers first published as *L'Avenir de l'homme* (Paris: Editions du Seuil, 1959).

19. The utopian ideal and modern opposition to it are thoroughly discussed by the American political scientist George Kateb in *Utopia and Its Enemies* (New York: Free Press, 1963). He pays particular attention to the unpalatable utopia of the father of programmed instruction, the Harvard psychologist Burrhus Frederick Skinner, *Walden Two* (New York: Macmillan, 1948pb); this is a self-sufficient community run on such behavioristic principles that no member would even think of asocial activity and all are free and creatively happy. In a more recent work, *Beyond Freedom and Dignity* (New York: Knopf, 1971), Skinner again argues that we can preserve civilization only by abandoning our cherished reverence for individual freedom in favor of "operant conditioning" for altruism.

20. Samuel Butler (1835–1902), whose most famous work is his autobiographical novel *The Way of All Flesh* (1903), is not to be confounded with another English satirist, the Restoration poet Samuel Butler (1612–1680). Kateb (see n. 18 above) draws a direct line from *Erewhon* and its sequel, *Erewhon Revisited* (1901), to such 20th-century antiutopian works as Capek's *R.U.R.* (cited in chap. 4) and Eugene Zamyatin's *We* (New York: Dutton, 1924bp), which stress the irreconcilable opposition between the natural and the mechanical.

21. Franz Kafka (1883–1924) wrote a number of works in this vein, most of them published posthumously, of which the best known is *The Trial* (1925pb).

22. Edward Morgan Forster's short story, "The Machine Stops," appears in *The Eternal Moment and Other Stories* (London: Sidgwick & Jackson, 1928pb).

23. In the 20th century, science fiction ("sci-fi" to the trade) has dominated utopian writing, especially in the work of Herbert George Wells (1866–1946) and his followers, including the brilliant Soviet writer Ivan Antonovich Efremov, some of whose works are beginning to be translated into English. However, it

must not be thought that utopia and science fiction are incompatible. For instance, Bellamy's book predicts closed-circuit broadcasting, including a touching paean to the uses that the 19th-century protagonist sees the Americans of the year 2000 making of the invention: "If we could have devised an arrangement for providing everybody with music in their homes, perfect in quality, unlimited in quantity, suited to every mood, and beginning and ceasing at will, we should have considered the limit of human felicity already attained, and ceased to strive for further improvements." One work of that master of science fiction, the patriotic Jules Verne (1828–1905), *Les cinq cents millions de la bégum* (1879), describes a model town run on humanitarian principles by its French founder and contrasts it with the authoritarian society in the nearby community founded by his German adversary.

24. The citation, which Huxley used as an epigraph to *Brave New World* (London: Chatto & Windus, 1932[pb]), is from the essay on "Democracy, Socialism and Theocracy," by Nicholas Berdyaev, in *The End of Our Time* (London: Sheed, 1933). The phrase "brave new world" itself is from act 5 of Shakespeare's *The Tempest.*

25. Milovan Djilas organized, with Josip Broz Tito, the Yugoslav communist party and provided the ideological basis for "Titoism," the policy of running a communist state independently of the USSR, which led to Yugoslavia's break with the Communist Information Bureau in 1948. (In 1956, as a gesture of reconciliation, Cominform was dissolved.) Following the forcible subjugation of Hungary by the USSR, which Djilas opposed, he was imprisoned; his term was prolonged upon the publication, in America, of *The New Class: An Analysis of the Communist System* (New York: Praeger, 1957[pb]). He was conditionally released in 1961, but rearrested after publishing *Conversations with Stalin* (New York: Harcourt, Brace & World, 1962[pb]) on the grounds of having divulged official secrets. He was released in 1966. Following the forcible subjugation of Czechoslovakia by the USSR, which both Djilas and Tito opposed, Djilas wrote *The Unperfect Society* (New York: Harcourt, Brace & World, 1969), in which he argued that Soviet totalitarianism is not an aberration of Marxism but its inevitable result, and that communism is doomed to disappear because the professional classes on which it depends are pressing for a more active and flexible society.

26. The proceedings of the Encyclopaedia Britannica Conference on The Technological Order are published in *Technology and Culture* 3(Fall 1962): 381–658. Ellul's contribution includes the discourse on the ambiguity of technique mentioned above (n. 15); this extract from the discussion is on p. 446.

27. Anatoli A. Zvorykin served as senior member of the staffs of both the Institute of Philosophy and the Institute of History of the USSR Academy of Sciences, and coauthored (with N. I. Osmova, V. I. Chernyshev, and S. V. Shukhardin) the standard Soviet *History of Technology* (Moscow: Izdatelstvo Socialno-Ekonomicheskoi Literatury, 1962), in Russian, also available in German as *Geschichte der Technik* (Leipzig: VEB Buchverlag, 1962). An earlier glimpse of Zvorykin's views is available in his article, "The History of Technology as a Science and as a Branch of Learning: A Soviet View," *Technology and Culture* 2(1961):1–4.

28. See, for instance, L. J. Halle, *The Society of Man* (London: Chatto & Windus, 1965[pb]), chaps. 5 and 6.

29. The argument regarding the role of science in production is taken up in the book by a team (at the Institute for the History of Science and Technology

in Moscow) directed by S. V. Shukhardin, *Sovremennaya Nauchno-Tekhnicheskaya Revolyutsia (*Moscow: Izdatelstvo Nauka, 1967, 1970); an English translation, *The Contemporary Scientific and Technological Revolution*, is to be published by San Francisco Press, 1973.

30. Radovan Richta, ed., *Civilizace na rozcestí* (Prague: Nakladatelství Svoboda, 1967); English translation and enlarged edition, *Civilization at the Crossroads: Social and Human Implications of the Scientific and Technological Revolution* (White Plains, N.Y.: International Arts and Sciences Press, 1969).

Notes to Chapter 7

1. For a general reference, see U.S. Department of Agriculture Miscellaneous Publication No. 15, A. C. True, *A History of Agricultural Extension Work in the United States, 1785–1923* (Washington, D.C.: U.S. Government Printing Office, 1928), written by a USDA "Specialist in States Relations Work."

2. The standard biography of Knapp is by J. C. Bailey, *Seaman A. Knapp: Schoolmaster of American Agriculture* (New York: Columbia University Press, 1945).

3. The U.S. Department of Agriculture, headed by a commissioner, was created in 1862. It was given cabinet status under a secretary of agriculture in 1889.

4. The first agricultural experiment station in America (in Connecticut) antedated the Hatch Act (1887) by a decade and was based on European models; but such efforts remained small potatoes until the infusion of federal subsidies. The bill was drafted by Knapp and others and introduced by the Missouri congressman William Henry Hatch (1833–1896).

5. Agricultural extension worked closely during 1917–1918 with the U.S. Food Administration, then headed by the future President Herbert Clark Hoover (1874–1964).

6. The 4-H Clubs (each member pledges his head, heart, hands, and health to the program) were started in 1914 by Jessie Field Shambaugh (1881–1971) to help young people (aged 10–21) in a practical way to acquire skills in agriculture and home economics and to foster high ideals of civic responsibility and international understanding. The movement has spread to over 75 countries. In the USA, local groups are often guided by the county agents.

7. For instance, a report by the dean of agriculture at the University of Ibadan describes efforts made in Nigeria to introduce an extension service on the American pattern and speaks wistfully of the responsibility that local farmers' organizations take for agricultural extension in, say, New York State. See V. A. Oyenuga, *Agriculture in Nigeria* (Rome: U.N. Food and Agriculture Organization, 1967), p. 291.

8. Of the massive literature of Nazi abuses, the present account relies mainly on the official transcript of the trials held by American prosecutors during the years 1946–1949 at Nuremberg *after* the International Military Tribunal had tried the major war criminals, 1945–1946, especially vol. 5 of the *Trials of War Criminals Before the Nuremberg Military Tribunals* (Washington, D.C., U.S. Government Printing Office, 1950). Among students of the question of national responsibility for Nazi crimes, the most thoughtful is the German-Swiss historian M. G. Steinert, author of *Hitlers Krieg gegen die Deutschen* (Düsseldorf and

Vienna: Econ Verlag, 1970) and *L'Allemagne nationale-socialiste* (Paris: Editions Richelieu, 1972).

9. The silence of religious leaders in the face of their knowledge of mass exterminations has been dramatized by Rolf Hochhuth's 1962 play *Der Stellvertreter* ("The Deputy," that is, God's vicar on earth), in which Pope Pius XII (1876–1958) was taken to task for failing to intercede with the Nazis against the exterminations.

10. The Treblinka data are taken from the carefully researched book by Jean-François Steiner, *Treblinka* (London: Weidenfeld and Nicolson, 1967pb; first published in Paris in 1966); the cited portion is on pp. 48–49.

11. Auschwitz ultimately became a sizable complex of 40 square kilometers containing three camps: the labor camp at Auschwitz, the extermination camp near the birch forest of Birkenau (Brzezinka), and the "external" camp near Monowitz (Monowice). The last comprised agricultural and industrial installations that had sprung up near this ready source of cheap labor; I. G. Farbenindustrie employed thousands in its synthetic-rubber and gasoline plants at Monowitz, as did other big firms, for instance Siemens & Schuckert at nearby Bobrek and some of the coal mines of Upper Silesia. The wages paid to the slave laborers naturally remained in the SS treasury.

12. Gerald Rechtlinger, *The Final Solution* (London: Valentine, Mitchell, 1953), p. 250.

13. Originally published in French, Saul Friedländer's *Kurt Gerstein ou l'ambiguité du bien* (Paris: Casterman, 1967) appeared in English as *Kurt Gerstein: The Ambiguity of Good* (New York: Knopf, 1969).

14. For the story of Allied indifference to the reports about the exterminations, see A. D. Morse, *While Six Million Died* (New York: Random House, 1967pb).

15. On this point it is worth noting that the principal reproach levelled against the American physicist J. Robert Oppenheimer (1904–1964) in the notorious 1954 hearings (as a result of which he was denied further access to secret information) was that as a government consultant he "did not show the enthusiastic support for the Super program [the development of an early version of the hydrogen bomb] which might have been expected of the chief adviser to the Government under the circumstances," as if enthusiasm were a prerequisite for technical consultants.

16. See, for instance, a prominent engineer's attempt to reduce such factors to a common denominator (money): Chauncey Starr, "Social Benefit versus Technological Risk," *Science* 165(1969):1232–1238.

17. Adolf Augustus Berle, Jr. (1894–1971), served as member of President Franklin Roosevelt's famed group of advisors nicknamed the Brain Trust and later as assistant secretary of state and ambassador. He was a well-known writer of books on political and economic theory, of which the study he coauthored with Gardiner C. Means, *The Modern Corporation and Private Property* (New York: Macmillan, 1934pb), is considered a classic in its field. The list cited here is taken from a magazine article, "What GNP Doesn't Tell Us," *Saturday Review*, 31 August 1968, p. 10.

18. The original version of the Hippocratic Oath is as follows:

"I swear by Apollo the physician, by Aesculapius, by Hygeia, Panacea, and all the gods and goddesses, that, according to my best ability and judgment, I will keep this oath and stipulation; to reckon him who taught me this art equally dear to me as my parents; to share my substance with him and relieve his necessities if required; to regard his offspring as my own brothers and to teach them this art if they shall wish to learn it, without fee or stipulation, and that by precept, oral teaching, and every other mode of instruction I will impart a knowledge of the art to my own sons and to those of my teachers, and to disciples bound by a stipulation and oath, according to the laws of medicine, but to no others.

"I will follow that method of treatment which, according to my ability and judgment, I consider for the benefit of my patients, and abstain from whatever is deleterious and mischievous. I will give no deadly medicine to anyone if asked, nor suggest any such counsel; furthermore, I will not give to a woman an instrument to produce abortion.

"With purity and with holiness I will pass my life and practice my art. I will not cut a person who is suffering with a stone, but will leave that to be done by practitioners of such work. Into whatever houses I enter I will go into them for the benefit of the sick and will abstain from every voluntary act of mischief and corruption, and, further, from the seduction of females or males, bond or free.

"Whatever in connection with my professional practice, or not in connection with it, I may see or hear in the lives of men which ought not to be spoken abroad, I will not divulge, as reckoning that all such should be kept secret.

"While I continue to keep this oath inviolate, may it be granted to me to enjoy life and the practice of my art, respected always by all men, but should I trespass and violate this oath, may the reverse be my lot."

A modified form was adopted by the World Medical Association in 1948 and is known as the Declaration of Geneva; the Engineer's Oath suggested here is adapted from that version.

A number of prominent engineers who read the section containing An Engineer's Hippocratic Oath expressed various degrees of agreement with the concept. Many were enthusiastically in favor. Others mentioned difficulties arising from the division of loyalties between one's employer and the larger society, the fact that many engineers worked in defense industries, and the fact that, as one (Harvey Brooks) put it, "few of these general principles are very helpful in resolving the practical moral decisions with which the engineer is faced in real life." He added, "There is also the question of when he has a right, or possibly a duty, to withhold his skills from use for purposes which the society to which he belongs has democratically determined to be valid." If the present contribution engenders widespread discussion of these points, an important purpose will have been served.

19. The sayings attributed to Georges Clemenceau (1841–1929), twice prime minister of France, is "War is too important to be left to the generals."

Notes to Chapter 8

1. Cited in G. A. Codding, Jr., *The International Telecommunications Union* (Leiden: E. J. Brill, 1952). The chairman was the American lawyer and broadcasting executive, Charles Ruthven Denny, Jr., then head of the U.S. Federal Communications Commission.

2. Examples of each of these categories of the literature of development and international aid are Barbara Ward, *The Rich Nations and the Poor Nations*

(New York: W. W. Norton, 1962pb), by an English writer on international relations; L. J. Zimmerman, *Poor Lands, Rich Lands: The Widening Gap* (New York: Random House, 1965pb), by a Dutch economist; and Raymond Aron, ed., *World Technology and Human Destiny* (Ann Arbor: University of Michigan Press, 1963), a transcript (first published in French in 1960) of a conference of prominent thinkers in several fields organized at Basel-Rheinfelden by a French philosopher. There is also an international association, the Society for International Development, with members in 120 countries, which publishes a quarterly *International Development Review* and a monthly newsletter, the *Survey of International Development;* and there are national organizations and publications in many countries. For instance, the American Society of Engineering Education has an active International Division, and countless other technical societies have formal ties with counterparts in several countries, often under the auspices of the United Nations Educational, Scientific and Cultural Organization (UNESCO).

3. H. E. Hoelscher and M. C. Hawk, eds., *Industrialization and Development* (San Francisco: San Francisco Press, 1969).

4. N. W. Chamberlain, "Training with Human Capital," in D. L. Spencer and Alexander Woroniak, eds., *The Transfer of Technology to Developing Countries* (New York: Praeger, 1967), chap. 6.

5. Margaret Mead, ed., *Cultural Patterns and Technical Change* (New York: Columbia University Press, 1955pb).

6. During the postwar years 1945–1961, the USA expended about $33 billion in nonmilitary aid. Since 1961, the Organization for Economic Cooperation and Development (OECD) has accounted for 90 percent of the aid given to developing countries by the noncommunist world. The principal donors are the USA, UK, France, West Germany, and Japan; the USA has been contributing the lion's share, although on a basis of percentage of national income France has given three times as much (about 2.7 percent to the USA's 0.9 percent), notably to its former colonies. Much of the American aid is channeled through the U.S. Department's Agency for International Development (AID), which stresses long-term goals such as technical training; for an evaluation of that particular activity, see A. E. Gollin, *Education for National Development* (New York: Praeger, 1969), as well as the book cited in n. 3 above.

7. In addition to a statistical survey, the study also contains a critical review of the literature on the performance of students who return to their countries. Charles Susskind and Lynn Schell, *Exporting Technical Education: A Survey and Case Study of Foreign Professionals with U.S. Graduate Degrees* (New York: Institute of International Education, 1968).

8. On this point, a French economist commenting on the absence of a flow of French scholars to the USA has observed, not without irony: "Many things are cheaper in the USA (gasoline, clothing, refrigerators, etc.); but the things the Frenchman cares for are relatively more costly (health, care and schooling of children, domestic services, repair jobs, vacations, good restaurants). On the other hand, the Frenchman may not greatly appreciate a perfect lawn in front of the house, a magic shower, and packaged and sliced bread. Finally, the French intellectual abominates do-it-yourself (and the bondage of dishwashing and laundrying)." Robert Masse, "Le cas français," in Walter Adams and Henri Rieben, eds., *L'Exode des cerveaux* (Lausanne: Centre de Recherches Européennes, 1968), chap.

10. An English version is Walter Adams, ed., *The Brain Drain* (New York: Macmillan, 1968).

9. A readily accessible expression of this viewpoint is an article by Kingsley Davis, "Population Policy: Will Current Programs Succeed?" *Science* 158(1967): 730–734.

10. See, for instance, P. R. Ehrlich, *The Population Bomb* (New York: Ballantine, 1968pb).

11. T. R. Malthus, *An Essay on the Principle of Population as It Affects the Future Improvement of Society* (London: J. Johnson, 1798). Malthus is also important in the history of economics, notably for his concepts of "effective demand" and of the balance between a nation's "power to produce and the will to consume," which to some extent anticipate the work of Lord Keynes. See also Vito Volterra, *Variazioni e fluttuazioni del numero d'individui in specie animali conviventi* (Venice: Regio Comitato Talassografico Italiano, 1927).

12. J. B. Wiesner and H. F. York, "National Security and the Nuclear-Test Ban," *Scientific American* 211(April 1964):27–35. "It is our considered professional judgment," they write, "that this dilemma has no technical solution." See also H. F. York, *Race to Oblivion: A Participant's View of the Arms Race* (New York: Simon and Schuster, 1970).

13. Garrett Harding, "The Tragedy of the Commons," *Science* 162(1968):1243–1248. The extreme view that science and technology can contribute relatively little is contested in the riposte by B. L. Crowe, "The Tragedy of the Commons Revisited," *Science* 166(1969):1103–1107.

14. Sebastian de Grazia, *Of Time, Work and Leisure* (New York: The Twentieth Century Fund, 1962pb). Other works touching on this subject are R. W. Kleemeier, ed., *Aging and Leisure: A Research Perspective into the Meaningful Use of Time* (New York: Oxford University Press, 1961), to which De Grazia contributed a chapter; the profusely illustrated book by Foster Rhea Dulles, *A History of Recreation: America Learns to Play* (New York: Appleton-Century-Crofts, 1960, 1965pb); Eric Larrabee and Rolf Meyerson, eds., *Mass Leisure* (Glencoe, Ill.: The Free Press, 1959); and of course the previously mentioned work of George Kateb, *Utopia and Its Enemies* (New York: Free Press, 1963).

15. Dennis Gabor, *Inventing the Future* (London: Seeker and Warburg, 1963pb; New York: Knopf, 1964). Gabor received the Nobel prize in physics in 1971 for his work, in the late 1940s, on a lensless method of producing three-dimensional displays that remained of largely theoretical interest until well after the invention of the laser in 1964 (see end of chap. 3). For Gabor's contributions to communication theory, see n. 9 of chap. 4. Gabor has carried his social ideas further in two other books, *Innovations, Scientific, Technological and Social* (New York: Oxford University Press, 1970); and *The Mature Society* (New York: Praeger, 1972). He sees technology as once again having a clear-set aim: to develop in a way that feeds only on inexhaustible or self-renewing resources, so as to make a final equilibrium state possible that is bearable and can last more than a few hundred years. For a tentative start in that direction, see a report of the Club of Rome's project on the predicament of mankind, D. L. Meadows et al., *The Limits of Growth* (New York: Universe Books, 1972).

16. The classic work on the relevance of societal institutions is by C. Northcote Parkinson, *Parkinson's Law* (Boston: Houghton Mifflin, 1957pb); wastefulness was also analyzed in a popular work by Vance Packard, *The Waste Makers* (New York: David McKay Co., 1960pb).

17. Gabor, *Inventing the Future*, p. 185.

18. The term "technology assessment" was coined in 1966 by Philip Yeager, counsel of the U.S. House of Representatives Committee on Science and Astronautics, the chairman of whose subcommittee on science, research, and development during the years 1963–1970, Congressman Emilio Quincy Daddario of Connecticut, was the first major figure to champion the concept.

19. Preoccupation with "national priorities" in the USA is also a relatively recent phenomenon. Much of the impetus has come from academic and professional circles; see, for instance, the proceedings of two symposia organized on the subject in 1970 by Stanford and by the Institute of Electrical and Electronics Engineers and combined into a single volume edited by Kan Chen, *National Priorities* (San Francisco: San Francisco Press, 1970), a politically well-balanced compendium that includes a pioneering quantitative study by the American economist Leonard A. Lecht, "National Priorities, Manpower Needs, and the Impact of Diminished Defense Purchases for Vietnam."

20. See, for instance, L. M. Branscomb, "Taming Technology," *Science* 171(1971):972–977.

21. Best known of the writings on this subject is the book by J. J. Servan-Schreiber, *Le défi americain* (Paris: Denoël, 1967), published in the USA as *The American Challenge* (New York: Atheneum, 1968).

22. Robert Gilpin: "Technological Strategies and National Purpose," *Science* 169(1970):441–448.

23. J. D. Carroll, "Participatory Technology," *Science* 171(1971):647–653.

24. See, for instance, L. H. Mayo, *Scientific Method, Adversarial System, and Technology Assessment* (Washington, D.C.: George Washington University, 1970). For a related concept, advocacy planning, see Paul Davidoff, "Advocacy and Pluralism in Planning," *Journal of the American Institute of Planners* 31(1965): 331–338; and L. R. Peattie, "Reflections on Advocacy Planning," *ibid.* 34(1968): 80–88. Among the unresolved questions of advocacy planning is the extent to which it might be useful to groups with neither economic nor political power (such as the poor), which lack the resources that can command technical expertise, insure legal access to information, guarantee protection against retribution, and generally provide the endurance without which few political actions can be effective.

25. John McPhee, *Encounters with the Archdruid* (New York: Farrar, Strauss & Giroux, 1971), contains narratives about conservationist David Brower and "three of his natural enemies" (including the U.S. Commissioner of Reclamation) that originally appeared in *The New Yorker* magazine.

26. Cost-effectiveness analysis, as formulated by C. J. Hitch, is a method of systems analysis derived from operations research, in the first instance to permit rational evaluation of alternatives available to defense planners in the USA. It is a special application of cost-benefit analysis, developed by Arthur Cecil Pigou (1877–1959) and his successors. Together with the planning method known as "program budgeting," the technique constitutes the planning-programming-budgeting system (PPBS), by which *substantive* planning ("planning by objectives") is linked to *fiscal* planning (budgeting), two functions hitherto usually separated. PPBS has gained gradual acceptance among government officials who must make allocative expenditure decisions (not only for defense but also for domestic purposes), and to some extent among their opposite numbers in private enterprise as well.

27. Arthur Pigou, *The Economics of Welfare* (4th ed.; New York; Macmillan, 1932). For an early attempt at government-sponsored technology assessment, see House Document 360 (75th Cong., 1st sess.), *Technological Trends and National Policy, Including the Social Implications of New Inventions,* Report of the Subcommittee on Technology to the National Resources Committee (Washington, D.C.: U.S. Government Printing Office, 1937), which contains appraisals of future developments in nine areas of technology.

28. Both are included in a list of 62 major breakthroughs since 1900: K. W. Deutsch, John Platt, and Dieter Senghaas, "Conditions Favoring Major Advances in Social Science," *Science* 171(1971):450–459.

29. *Technology: Process of Assessment and Choice,* Report of the National Academy of Sciences (Washington, D.C.: U.S. Government Printing Office, 1969), pp. 81–82.

30. *Ibid.*, p. 3.

31. *Ibid.*, p. 6.

32. *A Study of Technology Assessment,* Report of the Committee on Public Engineering Policy, National Academy of Engineering (Washington, D.C.: U.S. Government Printing Office, 1969), p. 3.

33. *Technology: Process of Assessment and Choice,* p. 29.

34. K. J. Arrow, "A Difficulty in the Concept of Social Welfare," *Journal of Political Economy* 58(1950):328. Professor Arrow received the 1972 Nobel Prize for economic science.

35. Many psychologists and sociologists have studied alienation; readings can be found in Eric and Mary Josephson, eds., *Man Alone: Alienation in Modern Society* (New York: Dell Publishing Co., 1962pb), and in Gerald Sykes, ed., *Alienation: The Cultural Climate of Our Time* (New York: Braziller, 1964). Alienation among university students is the topic of a case study of a dozen Harvard liberal-arts undergraduates by Kenneth Keniston, *The Uncommitted: Alienated Youth in American Society* (New York: Dell Publishing Co., 1965pb).

36. Bruno Bettelheim, "Obsolete Youth," *Encounter* 33(September 1969):29; also available as a paperback (San Francisco: San Francisco Press, 1970).

37. Amitai Etzioni, *The Active Society: A Theory of Societal and Political Processes* (New York: Free Press, 1968).

Subject Index

Abacus, 48
Active society, 134, *135*, *152*
Advocacy, 128; in planning, *151*
Affluent Society, The, 85, *143*
Agency for International Development (AID), *149*
Aggression, 70, *140*
Agribusiness, 108
Agricultural extension, 105–108, 113
Agriculture, 2–14, 30, 62, 77, 105, 120, 123, *146*
Alienation, 132–134, *152*
American Challenge, The, 151
Analytical engine, 48, 50, *139*
Animal Farm, The, 96
Antiutopias, 93
Aquaculture, 64
Archeology, 77–78, *142*
L'Art éphèmère, 72
Artificial intelligence, 53–54, *139*
Artificial organs, 69
Atomic bomb, 61
Atomic energy, 61, 128
Auschwitz, 112–114, *147*
Automata, 43–44, 70
Automatic factory, 58
Automatic pilot, 39
Automatic Sequence Calculator, 49
Automation, 29, 39
Automobile, 21–22, 27
Aviation, 22

"Babi Yar," 110
Balloon frame, 26
Bauhaus, 71, *140*
Big Brother, 95–96, 100
Big science, 91, *144*
Binary system, 52
Bioengineering, 69
Biomedical sciences, 68–70
Birkenau, 112
Brain drain, 123, *149*

Note: Page numbers in italics signify entries in the notes.

Brave New World, 94, 96, *145*
Broadcasting, 24, 36
Brothers Karamazov, The, 95
Building technology, 24–26

CAI. *See* Computer-Assisted Instruction
Calculating machines, 48
Camera, 23
Candide, 48
Carbon dating, 78, *142*
Cardiography, 69
Categorical imperative, 116
Cat's whisker detector, 44–45
Chartism, 13
Cheaper by the Dozen, 138
China, 99
Cinematography, 71
Civilization and Its Discontents, 88
Civilization at the Crossroads, 98–99, *146*
Club of Rome, x
Coal mining, 8–9
Cominform, *145*
Communications, 23–24, 91, *137*
Communication theory, 43, 57, *140*
Communism, 13, 83, 96, 98, 100–101, *145*
Computer, 43–58, 70, 74, 77, 131, *139*; language for, 52
Computer-Assisted Instruction (CAI), 76–77
Concrete, 25–26
Controlled thermonuclear reactions, 62
Cooperative movement, 5, 12–13
Corn laws, 11
Cost-benefit analysis, 129
Cost-effectiveness analysis, *151*
Crystal Palace, 16, 26, 71, *140*
Cultural Patterns and Technical Change, 122
Cybernetics, 43–44, 57–58, *139*

Data processing, 39
Declaration of Independence, 12, 116
Decline of the West, The, 88
Deputy, The, 115
Deus ex machina, 70
Developing countries, 15, 62, 119–124, *148–149*
Dialectical materialism, 97; and dialectic, 97
Diesel engine, 22
Differential Analyzer, 49
Digital computer. *See* Computer
Diode, 44
Doublethink, 95

Ectogenesis, 94
Education, 91, *141*
Eiffel tower, 25–26
Electronic control, 55
Electronics, 35, 72–73, 139; industrial, 37–40; solid-state, 44–47, 51
Electron microscopes, 36, 69
Elevator, 26
Eminent domain, 104
Encephalography, 69
Energy, 59–62; conversion of, 17–19, 59
Engineering education, 30–32
Engineers and the Price System, The, 81
ENIAC, 50, 52
Enlightenment, 12
Environment, x
Erewhon, 94, *144*
Ersatz, 65
Esthetics, 87
Ethics, 103–118, *136. See also* Hippocratic oath
European Recovery Program, 122
Euthanasia, 109–115
Existentialism, 79
Experiments in Art and Technology, 72, *141*
Explorer, 46
Extermination camps, 108–115, *147*

Factory system, 3
Failure of Technology, The, 88
Feedback, 8, 37–38, 43, 55, 58, 129
Final Solution, 110, *147*
Fine arts, 70–74
Fission, 61
Food, 62–65
Foreign aid, 123
4-H Clubs, 108, *146*
Frankenstein, 53, *139*
Free Speech Movement, 133
Fuel cell, 60
Fusion, 61

Golem, 54–55, *139*

Governor, 43
Gramophone, 36. *See also* Phonograph
Great Eastern, 21
Great Exhibition, 16. *See also* Crystal Palace
Green revolution, 62–65

Hatch Act, 105–106, *146*
Hawthorne experiment, 29, 138
Hearing aid, 45, 69
Hippocratic oath, 105, 118, *147–148*
"Homage to New York," 72, *140*
Homestead Act, 113
L'Homme-machine, 92
Humanism, 79
Humanistic studies, 77
Hunger Fighters, The, 140
Hydrogen bomb, 61

ICOHTEC, *137*
Ideologies of technology, 79–101
Incrementalism, 94
Industrialization, 7, 10, 15–17, 28, 120, *136*
Industrialization and Development, 121, *149*
Industrial management, 28–30
Industrial research, 28, 35
Industrial Revolution, 2–14, 25, 33–35, 59, 71, 90, 93. *See also* Second Industrial Revolution
Information theory, 43, 57, *140*
Integrated circuits, 46
Internal combustion engine, 21–22
International Telecommunications Union, 120
International understanding, 119–124
Irish famine, 10

Job enlargement, 29

Kinetic art, 72

Land-grant colleges. *See* Morril Land-Grant College Act
Laser, 41, 69
League of Nations, 119
Leisure, 125–127, *150*
Limits of Growth, The, x, *150*
Linear prediction, 56
Linotype, 23
Looking Backward: 2000–1887, 80
Luddites, 11, *133, 136*

Macadam, 22
Machine tools, 16
Magnetic core, 47, 52
Managerial Revolution, The, 83–85, *143*
Managerial society, 81
Manifesto of the Communist Party, 13
Marketing, 28
Marshall plan. *See* European Recovery Program
Marxist views of technology, 97–101
Maser, 41
Mass production, 27–30
Materials, 65–68; processing of, 19–20, 65
Medicine, 118
Metallurgy, 34
Metronome, 73
Military-industrial complex, 87
Mobiles, 72
Modern Times, 28
Moonlighting, *126*
Morrill Land-Grant College Act, 31, 105
Morse code, 23, 57
Multiphasic health screening, 131
Musique concrète, 36, 72

Napoleonic wars, 7, 11
National Priorities, 151
Nazism, 108–115, *140, 143, 146*
New Class, The, 96, *145*
New Industrial State, The, 85–88, *143*
New Left, 133
Nineteen Eighty-Four, 95–96
Nuclear power, 19
Nuclear reactor, 61

Objectivity, 115–116
Obsolete Youth, 133, *152*
On-line, real-time control, 56
Open Society and Its Enemies, The, 94
Operations research, 70
Organization man, 92, *144*
Otto engine, 21
Overpopulation, 103, 124–125, *136, 140*

Pacemakers, 69
Painting, 71

Panharmonium, 73
Parkinson's Law, 126, *150*
Participatory technology, 128, *151*
Phonograph, 23, 36, 74
Photoelectricity, 60
Photography, 71
Planning-programming-budgeting system (PPBS), *151*
Plastics, 34, 65–68
Player piano, 73
Portland cement, 26
Possessed, The, 95
Postindustrial society, 15
PPBS. *See* Planning-programming-budgeting system
Prague Spring, 101
Prestressed concrete, 26
Prince, The, 95
Privacy, 58
Public health, 69
Punched cards, 48

Radar, 40–41, 46, 56, 69, *138*
Radiology, 69
Radiotelegraphy, 24, 35
Railroads, 9, 20, 25
Rationalizations, 28
Reinforced concrete, 26
Relay, 47
Republic, The, 95
Resources, x, 118
Run, Computer, Run, 76, *141*
R.U.R., 53, *144*

Satellites, 139
Science and Technology for Development, 120
SCUBA, 78
Second Industrial Revolution, 33–41, 48, 58, 69, 72, 85, 98–99, 132–133. *See also* Industrial Revolution
Semiconductors, 44, 47
Servomechanism, 38, 43, 55, 58, 70
Seven Wonders of the World, 25, 70
Sherman Antitrust Act, *143*
Sierra Club, 128
Slavery, 11, 14, 79, 86, 91, 116, *136*
Slide rule, 48
Smith-Lever Act, 107
Social engineering, 127
Socialism, 12–13, 28, 83–84, 99–100
Society for International Development, *149*
Sociology of Invention, 138, 142
Solid-state electronics, 44–47, 51
Sorcerer's Apprentice, 53
Soviet Union, 97–101, *145*
Sputnik, 45
SS (*Schutz-Staffel*), 109–114, *147*
Steamboat, 8
Steam engine, 3, 8–9, 17–18, *137*
Steam turbine, 18
Steel, 16, 19–20
Stored program, 51
Supersonic transport, 128
Syndicalism, 83
System theory, 70, 131

Technische Hochschule, 30–31, *136*
Technocracy, 82–83, 85, *142*
Technological Society, The, 89–93, 96, *144*
Technology assessment, 105, 127–132, *151–152*
Technology transfer, 123–124, *135, 137, 148*
Technostructure, 86–87
Telecommunications, *148*. *See also* Communications
Telegraph, 23–24. *See also* Radiotelegraphy
Telephone, 24, 36
Television, 36
Tempest, The, 95, 145
Textiles, 5–7, *136*
Theory of the Leisure Class, The, 82
Thermoelectricity, 60
Thermoluminescence dating, 78
"Tragedy of the Commons, The," 125, *150*
Transistor, 40, 45
Transport, 2, 21–22, 91
Treason of the Intellectuals, The, 88
Treblinka, 111, 114, *147*
Trial, The, 144
Triode, 36, 45–47
Turbine, 18, 60
Turing test, 54
Two cultures, 77, *141–142*
2001—Space Odyssey, 139
Typewriter, 23

UNESCO, 122, *137, 140, 149*
UNIDO, 122
United Nations, 119
UNIVAC, 51
Utilitarianism, 12, 131, *136*
Utopia, 93–94, *144*. *See also* Antiutopias

Vulcanization of rubber, 22

Wall, The, 111
West Point, 30, *138*
World Bank, *143*

Name Index

Adams, Walter, *149–150*
Aiken, Howard, 49–50
Albert, Prince, 16, *136*
Alexander II (emperor), 14
Allen, F. R., 142
Aquinas, St. Thomas, 93
Ardrey, Robert, *135, 140*
Arkwright, Sir Richard, 5, *136*
Armstrong, E. H., 36
Armytage, W. H. G., *143*
Aron, Raymond, *149*
Arrow, K. J., 129, *152*
Ashby, Sir Eric, *138*
Aspdin, Joseph, 26
Augustine, Saint, 93

Babbage, Charles, 48, 50
Bacon, Sir Francis, 3, 93
Baekeland, L. H., 34
Baeyer, Adolf von, 34–35
Bailey, J. C., *146*
Bardeen, John, 45
Barnett, H. G., *142*
Basov, N. G., 41
Baťa, Tomás, 137
Bayer, Friedrich, 34
Beethoven, Ludwig van, 73, 88
Behrens, Peter, 71
Bell, A. G., 24
Bellamy, Edward, 80–81, 85, 93, *145*
Benda, Julien, 88
Bendix, Reinhard, *143*
Benz, Karl, 22
Berdyaev, Nicholas, 95, *145*
Berle, A. A., Jr., 117, *147*
Bernstein, Jeremy, *139*
Bessemer, Sir Henry, 16, 20, 34, *137*
Bettelheim, Bruno, 133, *152*
Bismarck, Otto von, 138
Blackett, Lord P. M. S., *137*
Blackner, John, *136*
Blair, Eric, 45. *See also* Orwell, George

Note: Page numbers in italics signify entries in the notes.

Blake, William, 10
Bloch, Felix, 41
Boguslaw, Robert, *142*
Bond, F. A., *142*
Booth, Henry, 9
Brandt, Karl, 109–110
Branscomb, L. M., *151*
Brattain, W. H., 45
Brower, David, *151*
Brunel, I. K., 21
Burnham, James, 83–85, 89, *143*
Burroughs, W. S., 48
Bush, Vannevar, 49–50
Butler, Samuel, 94, *144*

Caecus, Appius Claudius, *137*
Calder, Alexander, 72
Čapek, Karel, 53, *144*
Carnegie, Andrew, 20
Carroll, J. D., *151*
Cartwright, Edmund, 5, *136*
Chamberlain, N. W., 121, *149*
Chaplin, Charlie, 28, 71
Chardin, Pierre Teilhard de, 93, *144*
Charlemagne (emperor), 70
Chen, Kan, *151*
Chernyshev, V. I., *145*
Clemenceau, Georges, 118, *148*
Codding, G. A., *148*
Commoner, Barry, x
Coupling, J. J., *139*. *See also* Pierce, J. R.
Crombie, A. C., *142*
Crompton, Samuel, 5, *136*
Crowe, B. L., *150*
Crozet, Claude, 30, *138*

Daddario, E. Q., *151*
Daimler, Gottlieb, 22
Darvall, F. O., *136*
Daumas, Maurice, *135*
Davidoff, Paul, *151*
Davis, Kingsley, 124, *150*
Day, B. H., *137*
Délibes, Léo, 53, *139*
Denny, C. R., *148*
Deutsch, K. W., *152*
Dexter, Lewis, *137*
Diesel, Rudolf, 22
Djilas, Milovan, 96, *145*
Dostoevsky, Fyodor, 95
Dugan, James, *137*
Dukas, Paul, 53
Dulles, F. R., *150*

Eckert, J. P., 50–51
Edison, T. A., 18, 23, 34–36
Efremov, I. A., *144*
Ehrlich, Paul, 34–35, *150*
Eiffel, Gustave, 25
Einstein, Albert, 61
Eisenberg, Leon, *140*
Eisenhower, D. D., 87, *143*
Ellul, Jacques, 89–92, 95–96, 99, 144–145
Elsner, Henry, 142
Engels, Friedrich, 13, 97–98
Etzioni, Amitai, 134, *152*
Evans, Oliver, 8

Finch, J. K., *136*
Fink, D. G., *139*
Fiore, Quentin, x
Fleming, J. A., 44
Flinn, A. D., 104
Floss, Herbert, 112–114
Ford, Henry, 27, 94
Forest, Lee de, 36
Forman, Sidney, *138*
Forster, E. M., 94, *144*
Fourastié, Jean, 93
Frankel, Charles, 75, 77, *141*
Franklin, Benjamin, *135*
Franz, Kurt, 111, 113–114
Freud, Sigmund, 88, 93
Friedländer, Saul, *147*
Friedmann, Georges, 93, *144*
Frisch, Ragnor, 93
Fulton, John, 8

Gabor, Dennis, 57, 126, *139–140*, *151*
Galbraith, J. K., 85–87, *143*
Galloway, B. T., 107
Gantt, H. L., *138*
Gerstein, Kurt, 114–115, *147*
Gilbreth, F. B., *138*
Gilbreth, L. M., *138*
Gilfillan, S. C., *138*, *142*
Gilpin, Robert, 127, *151*
Godwin, Mary, 53, *139*
Goethe, J. W. von, 53, *139*
Gollin, A. E., 149

Gompers, Samuel, 27
Goodyear, Charles, 22
Gottwald, Klement, *137*
Gould, J. M., *142*
Grazia, Sebastian de, 126, *150*
Gregory, Gene, *140*
Griffith, D. W., 71
Gropius, Walter, 71, *140*

Hall, A. R., *135*
Halle, L. J., *145*
Hammond, A. L., *142*
Harding, Garrett, 125, *150*
Hargreaves, James, 5, *136*
Hart, Hornell, *142*
Harun al-Rashid (caliph), 70
Hatch, W. H., *146*
Hawk, M. C., *149*
Hegel, G. W. F., 97
Heron, A. R., *138*
Hersey, John, 111
Hess, Rudolf, 112
Hill, Sir Rowland, *137*
Himmler, Heinrich, 112
Hindemith, Paul, 73
Hippocrates of Cos, 118
Hitch, C. J., *151*
Hitler, Adolf, 84, 95, 108, 110, 112
Hobsbawm, E. J., *136*
Hochhuth, Rolf, 115, *147*
Hoelscher, H. E., *149*
Hoess, Rudolf, 112–114
Hoffmann, E. T. A., 53, *139*
Hoffmann, Felix, *138*
Hofmann, A. W. von, *138*
Hollerith, Herman, 48, 50, *139*
Hollerith, Virginia, *139*
Holmyard, E. J., *135*
Hoover, H. C., *146*
Hulten, K. G. P., *140*
Huxley, Aldous, 94–97, *145*
Huxley, Julian, 94
Huxley, T. H., 94
Hyatt, J. W., 34, 65

Jacquard, J. M., 5, 48
John XXIII (pope), 116
Johnson, L. B., *141*
Joseph II (emperor), 14
Josephson, Eric, *152*
Josephson, Mary, *152*
Jünger, Ernst, *144*
Jünger, F. G., 88, *144*
Justinian (emperor), *137*

Kafka, Franz, 94, *144*
Kant, Immanuel, 116
Kateb, George, *144, 150*
Kay, John, 5, *136*
Keniston, Kenneth, *152*
Kennedy, J. F., 86, *143*
Keynes, Lord J. M., 85–86, 91, *150*
Kierkegaard, Søren, 79
Kistiakowsky, G. B., *143*
Kleemeier, R. W., *150*
Klüver, Billy, 72, *140*
Knapp, S. A., 105–106, 113, *146*
Kolmogorov, A. N., 56
Kranzberg, Melvin, *135*
Kruif, Paul de, *140*
Kubrick, Stanley, *139*

Langmuir, Irving, 35
Larrabee, Eric, *150*
Launay, Prosper de, 3
Lecht, L. A., *151*
Le Corbusier, 71
Leibniz, G. W. von, 48
Leland, H. M., 27
Lenin, Nikolai, 97–98
Leo XIII (pope), 116
Leonardo da Vinci, 70
Levin, R. E., *141*
Libby, W. F., *142*
Linden, Herbert, 109
Löw, Rabbi, 54
Long, Huey, 80
Lorenz, Konrad, 140
Ludd, Ned, 11, *136*

McAdam, J. L., 22
McDermott, John, *141*
Machiavelli, Niccolò, 95
McLuhan, Marshall, x
McPhee, John, *151*
Mälzel, J. N., 73
Malthus, T. R., 7, 125, *150*
Marconi, Guglielmo, 24
Marks, Sema, *141*
Marshall, G. C., 122
Marx, Karl, 10, 13, 80, 91, 93, 97–100
Mason, O. T., *136*

Masse, Robert, *149*
Mauchly, J. W., 50, 51
Mayo, Elton, 29, *138*
Mayo, L. H., *151*
Mead, Margaret, 122, *149*
Meadows, D. L., *150*
Means, G. C., *147*
Melville, Herman, 11
Mennecke, Fritz, 110
Mettrie, J. O. de la, 92
Meyerson, Rolf, *150*
Mies van der Rohe, Ludwig, 71
Mill, J. S., 7, *136*
Miller, D. C., *142*
Ming Dynasty, *137*
Mohammed Ali, 6, *136*
Moog, R. A., 73
Moore, G. E., *136*
More, Sir Thomas, 93
Morison, E. E., *137*
Morris, Desmond, *140*
Morris, William, 71, 94
Morse, A. D., *147*
Morse, S. F. B., 23, 57
Mournier, Emmanuel, 93
Mumford, Lewis, 89, *135*

Nader, Ralph, x
Napoleon (emperor), 6, 80, *138, 142*
Nervi, Luigi, 71
Neumann, John von, 51
Newcomen, Thomas, 8, 17
Newman, J. R., *139*
Newton, Isaac, 48
Niebuhr, Reinhold, 116
Nietzsche, F. W., 94
Nimkoff, M. F., *142*
Nott, Eliphalet, 106

Oakley, K. P., *135*
Oettinger, A. G., 76–77, *141*
Offenbach, Jacques, 53, *139*
Ogburn, W. F., 82, *142*
O'Meara, B. E., *142*
Oppenheimer, J. R., *147*
Orwell, George, 84, 94–96, *143*
Osmova, N. I., *145*
Otto, N. A., 21
Owen, Robert, 5, 12
Oyenuga, V. A., *146*

Packard, Vance, *150*
Parkinson, C. N., 126, *150*
Pascal, Blaise, 48
Paul, Lewis, 5, *136*
Paxton, Sir Joseph, 71, *140*
Peattie, L. R., *151*
Perkin, Sir W. H., 34, *138*
Peters, Armin, 115
Pierce, J. R., 43–44, *139*
Pigou, A. C., 129, *151–152*
Pius XI (pope), 116
Pius XII (pope), *147*
Plato, 93, 95
Platt, John, *152*
Pohl, Oswald, 114
Popov, A. S., *137*
Popper, Karl, 94
Porter, W. C., 107
Price, D. J. deSolla, *144*
Prokhorov, A. M., 41
Purcell, E. M., 41
Pursell, C. W., *135*

Rabelais, François, 93
Rankine, J. M., 18, *137*
Rathenau, Emil, 71
Rechtlinger, Gerald, *147*
Reich, Charles, x
Renault, Louis, 137
Rennie, George, 20
Ricardo, David, 7
Richta, Radovan, 101, *146*
Rieben, Henri, *149*
Rockefeller, J. D., 107
Roosevelt, F. D., 83, *143, 147*
Roszak, Theodor, x
Rumford, Count, 3

Saint-Simon, Comte de, 12
Samuel, A. L., 53
Savery, Thomas, 8
Schell, Lynn, *149*
Schmookler, Jacob, *142*
Scott, Howard, 83
Senghaas, Dieter, *152*
Servan-Schreiber, J. J., *151*
Shambaugh, J. F., *146*
Shannon, C. E., 57, *140*
Shelley, P. B., 53
Shockley, William, 45
Shukhardin, S. V., *145–146*

Singer, Charles, *135*
Skinner, B. F., *144*
Smeaton, John, 17, 26, 30
Smiles, Samuel, *136*
Smith, Adam, 7
Snow, Lord C. P., *141*
Solow, R. M., *143*
Spencer, D. L., *149*
Spengler, Oswald, 88, *143*
Spicer, E. H., *138*
Spillman, W. J., 107
Stakhanov, A. G., *138*
Stalin, Joseph, 84, 95
Starr, Chauncey, *147*
Steiner, Jean-François, 110, *147*
Steinert, M. G., *146*
Stendhal, *142*
Stephenson, George, 9
Stephenson, Robert, 9
Stibitz, G. R., 50
Stieglitz, Alfred, 71
Stockhausen, Karlheinz, 73
Stravinsky, Igor, 73
Susskind, Charles, *137*, *149*
Swift, J. D., 93
Sykes, Gerald, *152*

Taylor, F. W., 28, *138*
Tillich, P. J., 116
Tinguely, Jean, 72, *140*
Tito, J. B., *145*
Townes, C. H., 41
Toynbee, Arnold (1852–1883), 3
Toynbee, Arnold J. (b. 1889), 89
Trevithick, Richard, 8
Trotsky, Leon, 13
True, A. C., *146*
Turing, Alan, 51, 54, *139*

Ubbink, H. J., 115
Ussachevsky, Vladimir, 73

Van Allen, J. A., 46
Vanderpoel, Frank, 34
Varèse, Edgar, 73
Vaucanson, Jacques de, *139*
Veblen, T. B., 81–83, 85, *142*
Verne, Jules, *145*
Victoria, Queen, 16
Voltaire, 48
Volterra, Vito, 125, *150*

Ward, Lady Barbara, *135*, *148*
Watson-Watt, Sir R. A., 40
Watt, James, 8, 17, 37
Wells, H. G., *144*
Welsbach, C. A. von, 18
White, David, *137*
Whitney, Eli, 3, 5
Whyte, W. H., *144*
Wiener, Norbert, 43–44, 55–58, *139*
Wiesner, J. B., *150*
Wilkinson, John, 89
Williams, T. I., *135*
Wirth, Christian, 109, 113–114
Woroniak, Alexander, *149*
Wright, Frank Lloyd, 71
Wright, Orville, 22
Wright, Wilbur, 22
Wyatt, John, 5, *136*

Yeager, Philip, *151*
Yevtushenko, Evgeny, 110
York, H. F., *150*

Zamyatin, Eugene, 94, *144*
Zimmerman, L. J., *149*
Zvorykin, Anatoli, 97–99, *145*

THE JOHNS HOPKINS UNIVERSITY PRESS

This book was composed in Baskerville text and Typositor Baskerville display type by Monotype Composition Company from a design by Laurie Jewell. It was printed on Maple's 55-lb. Danforth and bound in Columbia Chambray cloth by The Maple Press Company.

Library of Congress Cataloging in Publication Data

Susskind, Charles.
Understanding technology.

Includes bibliographical references.
1. Technology. I. Title.
T47.S83 600 72-12344
ISBN 0-8018-1304-2